The Green Paradox

The Green Paradox

A Supply-Side Approach to Global Warming

Hans-Werner Sinn

The MIT Press
Cambridge, Massachusetts
London, England

For information on quantity discounts, email special_sales@mitpress.mit.edu.

Set in Sabon by Toppan Best-set Premedia Limited. Printed and bound in the United States of America.

Library of Congress Cataloging-in-Publication Data
Sinn, Hans-Werner.
The green paradox : a supply-side approach to global warming / Hans-Werner Sinn.
p. cm.
Includes bibliographical references and index.
ISBN 978-0-262-01668-0 (hardcover : alk. paper) 1. Carbon offsetting. 2. Supply-side economics. 3. Global warming. I. Title.
HC79.P55S574 2012
363.738'746—dc23

2011020801

10 9 8 7 6 5 4 3 2 1

for Hans Heinrich Nachtkamp, who taught me intertemporal
economics

Contents

Preface

To my surprise, my book *Das grüne Paradoxon*[1] triggered an international scientific debate and a swelling flow of scholarly papers. I am glad that this updated English edition has become available. I hope that a wider international audience will now have a chance to consider the issues the book raises. The English version has been condensed and focused on topics more likely to interest an international readership.

Non-European readers may forgive me for the fact that my European background shows through here and there. Europe is the center of a "green" policy thrust. My home country, Germany, is the world champion in solar power and biodiesel, number two in wind power, and one of the few countries adamantly determined to turn their backs on nuclear power. In Germany, the "green" idea has developed a momentum that reminds me of the rigor and zeal of the time, about 500 years ago, when Protestantism was born in the same country. Because I am a sober economist by training, the quasi-religious aspects that I occasionally detect here have made me a bit wary. But I can reassure the reader that in this book I have tried to be as general, unbiased, and international as I can. The international bent also explains why all numerical data are given in metric units, and why all tons are metric tons. (In many places, to make it easier for American readers, I have given the equivalent non-metric measures in brackets.)

My motivation for writing this book was a certain degree of frustration with the prevailing governmental policies for fighting global warming. Throughout Europe and elsewhere, particularly in California, politicians are keen to curb consumption of fossil fuels. They are busily promoting alternative energy, improved building insulation, and more efficient cars. They forbid citizens to use traditional light bulbs, force them to buy expensive "green" electricity and biofuels, impose emission constraints on car engines, subsidize electric cars, impose tough norms on the insulation of buildings (forcing homeowners to wrap their buildings in wadding), and frighten their citizens by announcing that they will come up with even tighter measures in the future. These programs cost hundreds of billions, yet in many cases they achieve little or nothing. The relentlessly rising curve of worldwide CO_2 output does not deign to honor these efforts with even the merest downturn.

This does not mean that mankind should give up the fight. In the first two chapters I show that we do indeed have a problem, and I discuss which energy options we may apply to solve it. However, the discussion, unavoidably and disappointingly, ends on a note of skepticism toward some of the technical fixes touted in the special reports that fill pages in our newspapers every other week.

In quantitative terms, the most important "green" energy source currently being developed is biofuels. I find biofuels problematic, if not downright dangerous, for the climate and for world peace. For reasons I spell out in chapter 3, replacing fossil carbon with biocarbon not only accelerates global warming but also deprives the world's poor of their food. It puts in our tanks what they would like to see on their tables. The link that has been re-established between the market for fossil fuels and the market for biocarbon is an unfortunate development of historical dimensions, one that risks pushing mankind back into the Malthusian trap.

But my concerns go further. I find the policies against global warming often naive and counterproductive, since they focus purely on demand, neglecting the supply side of the carbon markets. President Mahmoud Ahmadinejad of Iran, President Hugo Chávez of Venezuela, the Arabian oil sheikhs, Vladimir Putin's oligarchs, and all the coal barons of this world simply do not figure in the policy programs. However, these resource owners are the real climate makers. By bringing fossil carbon back into the carbon cycle by supplying it to the markets, thereby enlarging the stock of carbon dioxide in the atmosphere, they determine the speed of global warming; consequently, they hold the fate of humanity in their hands.

The resource owners regard the tightening of "green" policy measures with increasing concern, because they perceive them as what they are: a way of destroying their future markets. Quite understandably, they try to pre-empt the corresponding wealth losses by extracting and selling their fossil fuels before their markets disappear. That is the Green

Paradox: announced future reductions in carbon consumption may have the effect of accelerating climate change now. I suspect that one of the reasons why the prices of fossil fuels fell in real terms from 1980 to about 2000, even though China and India emerged as new consumers in the market, can be attributed to the "green" saber-rattling that occurred during that period. Resource owners simply hurried to secure their wealth by extracting their resources before the environmentalists could seize them.

Whether I am too pessimistic remains to be seen. The scientific debate will have to clarify that. But it seems to be clear that it is high time for environmental policy to shift its focus from the demand for fossil fuels to the supply of such fuels. Instead of mulling over for the thousandth time which technical fixes could be applied to reduce CO_2 emissions, we should turn to the question of how to induce resource owners to leave more carbon underground. Unfortunately, that goal will not be easy to achieve with the policy tools that the industrialized countries have at their disposal. Uncoordinated idiosyncratic measures by single countries or by groups of countries (such as the European Union) will achieve nothing, other than frightening the resource owners even more and inducing them to overextract.

However, the toolbox available to policy makers is not entirely empty. I argue in this book that only a swiftly introduced "Super-Kyoto" system, combining all consuming countries into a seamless demand cartel using a worldwide cap-and-trade system, will help. This system should be supported by the levying of source taxes on capital income to spoil the resource owners' appetite for financial assets.

So far, policy makers exhibit not the slightest glimmer of thinking about how they could influence the supply side of the carbon market. Hundreds of resolutions, laws, and promotion programs have been promulgated, all aimed at curbing the demand for fossil fuels, without one mention of the supply side. Half of the market for fossil fuels has simply been disregarded.

Until recently, even science hadn't really paid attention to the supply side. Models of long-term fossil-fuel extraction didn't concern themselves with the climate. Climate models, in turn, typically didn't concern themselves with the extraction of such resources. The few exceptions had to be searched for with a magnifying glass. These were theoretical models that never made it into numerical climate-simulation models, let alone to the public policy debate. Only recently have more scientists, including one from the Intergovernmental Panel on Climate Change,[2] begun to explicitly model the supply side numerically, joining in the Green Paradox debate.

Precisely because I consider climate change one of the greatest problems humanity faces, I find this neglect profoundly disquieting. Thus, I hope that policy makers will read this book. Those who have learned to focus on the supply-side effects of their policies will shed their illusions and will support a climate policy that offers better chances of staving off disaster.

In writing this book I enjoyed help from a number of members of the Ifo Institute and the Center for Economic Studies at the University of Munich. A first translation of the second German edition was prepared by Julio Saavedra, who also helped me polish my style when I made a condensed and updated English version out of it. Occasional help with the English also came from Paul Kremmel. Christian Beermann, Petra Braitacher, Max von Ehrlich, Mark Gronwald, Darko Jus, Wolfgang Meister, Johannes Pfeiffer, Tilman Rave, Luise Röpke, Johann Wacker-bauer, and, above all, Hans-Dieter Karl supported me in data mining, in searches of the literature, and in various calculations. The graphs were prepared by Christoph Zeiner with the help of Jana Lippelt. Martin Faulstich, head of the Wissenschaftszentrum Straubing and chairman of Germany's Sachverständigenrat für Umweltfragen (Advisory Council on the Environment), read the entire German manuscript and gave me very useful advice. Maximilian Auffhammer of the University of California at Berkeley also made useful comments. Knut Borchardt, my admired senior colleague on the faculty of economics at the University of Munich, provided me with insightful comments on the history of industrialization. Finally, three anonymous referees consulted by the MIT Press made very valuable suggestions for further improvement of the manuscript. I thank all of them for their generous support.

The German predecessor of this book was dedicated to Sascha Becker, Helge Berger, Marko Köthenbürger, Kai Konrad, Ronnie Schöb, Marcel Thum, Alfons Weichenrieder, and Frank Westermann, former students with whom I had been able, over the previous decades, to gain some of the knowledge that has gone into this book. I dedicate this version to my former professor and thesis supervisor Hans Heinrich Nachtkamp on the occasion of his eightieth birthday. He taught me intertemporal economics and dynamic optimization about 35 years ago. After so many years and quite a number of complicated mathematical papers on intertemporal topics, I finally dared to talk about intertemporal economics in ordinary language that every educated person should be able to understand, regardless of his or her field of specialization, hoping that my fellow colleagues will not think this is non-science simply because the equations have been turned into words. I can assure them that this was more difficult than doing it the other way round.

Munich, January 2011

to the residents of Novosibirsk, with a suggestion that they start growing palm trees

1

Why the Earth Is Getting Warmer

Just a tiny little bit

Carbon dioxide (CO_2) is a non-toxic gas. It is contained in every carbonated drink, and its sparkle feels refreshing. But it also strikes fear in us, because ever-larger amounts of it are being released into the atmosphere, accelerating the greenhouse effect. Akin to the glass panels in a greenhouse, CO_2 traps sunlight and thereby warms the planet.

Burning the carbon contained in oil, coal, natural gas, wood, and other organic matter produces carbon dioxide. The fats, carbohydrates, and proteins burned chemically by living organisms also contain carbon.

It is striking how little carbon dioxide the atmosphere contains. It accounts for barely 0.038 percent of the atmosphere. Chemists refer to this as 380 ppm, with ppm standing for parts per million. Before industrialization, the CO_2 concentration was 280 ppm. And, by the way, any gas spreads out in the atmosphere in such a way that its molecules are separated equally from each other when the air pressure is the same. The volume ratio thus always corresponds to the ratio of the number of molecules.[1] Owing to the differing weight of the respective molecules, however, the weight ratios do not correspond to the volume ratios. CO_2 is a pretty heavy gas, and without constant movement of the air it would concentrate near the ground.

Oxygen and nitrogen constitute 97 percent of the atmosphere. Oxygen accounts for 21 percent; nitrogen accounts for 76 percent. The rest consists of approximately 2.5 percent water vapor and numerous trace gases, of which CO_2, at 380 ppm, is the most important for climate. The second most important is methane (produced by the decay of plant matter in the absence of oxygen, for instance in the stomachs of cattle); it accounts for 1.8 ppm. Greenhouse gases, strictly defined, include carbon dioxide, methane, nitrous oxide, and other trace gases. A broader definition includes water vapor.

Water vapor plays a very significant role in the greenhouse effect. It is usually present as an invisible, wholly diluted gas, but it can quickly condense at lower pressures or temperatures and turn into clouds, rain, or snow. Its concentration varies greatly. The greenhouse effect occurs only when water vapor is in its uncondensed, invisible form. As a rule, 96 percent of the water in the atmosphere consists of water vapor. The remaining 4 percent is in the form of water droplets and ice crystals in clouds, rain, and fresh snow.[2]

Greenhouse gases are, in fact, not a problem but a boon for mankind. As is often the case, it all depends on the right dose. If there were no strictly defined greenhouse gases and no water as vapor or clouds in the air, the atmosphere would consist exclusively of oxygen and nitrogen, which it nearly does anyway, and the planet would be barely inhabitable, because the average temperature at ground level would be −6°C (21°F). (It would be much colder if there were water in the form of vapor and clouds in the air but none of the strictly defined greenhouse gases, as the cloud cover would be so thick that little sunshine would reach the ground. This will be explained below.) At present, the average temperature at ground level is 14.5°C (58.1°F), whereas in pre-industrial times it was 13.5°C (56°F). Therefore, the greenhouse gases, including water vapor, caused an increase in the ambient temperature of about 20°C (36°F).[3]

In this light, we can count ourselves fortunate that greenhouse gases exist at all. It was these gases that made life as we know it possible. A temperature of 14.5°C doesn't sound very comfortable, but in fact it is quite acceptable if one considers that it is an average that encompasses the polar caps and the tropics, winter and summer, and day and night. This is a temperature at which both people and nature feel comfortable, because evolution made us for it. In the past million years of evolution, average temperatures were around 11°C (52°F), rising by 4°C (7.2°F) during interglacial warm periods and falling by 2°C (3.6°F) during ice ages. Plants and animals can cope with changes of this magnitude because they can move back and forth between cold and warm regions.

During the last ice age, which ended 18,000 years ago, the average temperature was about 5.5°C (10°F) lower than today's. No one lived in Europe north of the Alps, as practically the entire region was buried under an ice layer.[4] In Germany, which now has an average temperature of 9°C (48°F), the average temperature was approximately –4°C (25°F).[5] However, our ancestors found temperate areas in Africa, in India, in Australia, and around the Mediterranean. The African tropics, which now boast an average temperature of 26°C (79°F), had an average temperature of about 21°C (70°F) back then—a level that nowadays can be found in northern Egypt, Texas, or southern Italy. Southern Italy then had an average temperature similar to Germany's today—from 8°C to 10°C lower than today's average.[6]

But a boon can turn into a bane if the concentration of greenhouse gases increases as a result of mankind's activities, because the climate reacts with extraordinary sensitivity to such gases. If the current concentration (0.038 percent), plus water vapor, can raise temperatures by 20°C (36°F), an uncontrolled increase can quickly turn into a calamity. God preserve us from conditions such as those on Venus, whose atmosphere consists mainly of carbon dioxide and water vapor. The greenhouse effect there has brought about temperatures of 525°C (977°F), rendering neither life nor love possible. Lovers are advised to keep their distance from Venus.

The greenhouse effect

Behind the public pronouncements on the greenhouse effect lie sound theories and vast records of measurements and observations. As these are shared by practically every leading climate scientist, there can be little doubt about the greenhouse basics, despite some irritating public debates in recent years. The first studies of the greenhouse effect were conducted in the nineteenth century. A veritable flood of scientific publications on the subject are now available.[7] The recognized authority on interpretation of the data and application of the associated theories is the Intergovernmental Panel on Climate Change (IPCC), a network of about 2,500 researchers that monitors climate change and publishes regularly updated reports on the subject.[8]

Climate research starts from the fact that a gas that consists of at least three atoms acts as a filter, absorbing certain wavelengths in the infrared, growing warm, and passing this warmth on to the gases surrounding it. The energy contained in the sunlight radiated by the Earth back to space, mainly in infrared frequencies, plays a major role in this. Three-atom greenhouse gases include carbon dioxide (CO_2), water vapor (H_2O), nitrous oxide (N_2O), and ozone (O_3). Methane has the chemical formula CH_4, which means it consists of five atoms. The gases grouped under the name "chlorofluorocarbons" (abbreviated CFCs) have at least six atoms. Akin to a color filter that absorbs a certain spectrum and so tints everything in a certain hue, the greenhouse gases absorb certain spectral colors. Oxygen (O_2) and nitrogen (N_2) don't produce any greenhouse effect, as each of their molecules contains only two atoms.

Sunlight has a wide color spectrum that contains particularly high levels of energy in the shorter (blue) wavelengths. It reaches the surface of the Earth practically unhindered, warms it, and turns into infrared light, which is then reflected by the Earth. We can't see infrared light, but we can feel its warmth. The police are very fond of shooting pictures in infrared, and we can do that ourselves in order to find out where warmth leaks out of our houses. A significant portion of the infrared light reflected by the Earth is absorbed by the greenhouse gases, converted into heat, and thus prevented from being expelled into space. This keeps our planet warm.

It keeps it warm, but it doesn't make it ever warmer. In theory, the Earth has a stable average temperature whose level depends on greenhouse-gas concentrations and other factors. When we say that the temperature is stable, we don't mean that it is constant; we mean that after an external disturbance, such as a change in solar radiation or a displacement of the continents, it reverts to a new equilibrium. An egg carried in a spoon is in a stable position. Although it rocks to and fro, it returns to stillness once the person carrying the spoon stops moving. If, however, one places the egg on the back of the spoon, it will be in an unstable position, and the tiniest movement will send it tumbling down. The temperature of our planet will not swing explosively if a change in solar radiation occurs, but it will experience minor swings that dampen down with time, tending toward an equilibrium. Fortunately, temperature changes caused by external factors don't build up over time. If the temperature were not stable, life would not be possible on our planet, because the many disturbances during its history would have turned it alternatively into a frozen waste and then a stifling desert.

The temperature remains stable because the more energy the Earth absorbs, the more it radiates back into space. If external factors make the planet warmer than what corresponds to its stable temperature, it radiates more energy into space and so the increase in temperature is slowed. Conversely, when an external factor makes it cooler, the planet radiates less energy than it receives from the sun, which slows the decline of temperature. It is like a light bulb. The current flowing into it doesn't make the filament shine ever more brightly; its brightness is determined by an equilibrium between the amount of energy being dissipated as light and the amount of electrical energy flowing in.

Averaging out all its regions, winter and summer, day and night, the Earth, including its enveloping atmosphere, receives an amount of energy amounting to 343 watts per square meter (32 watts per square foot). It must then be warm enough to radiate exactly this amount of energy back into space. If the atmosphere contained only oxygen and nitrogen, and neither water vapor nor clouds, nor carbon dioxide, nor any of the other greenhouse gases, the air and the surface would reflect 55 watts back to space immediately, so that 288 watts would remain to warm the planet's surface and its air. The temperature on the surface would then stabilize at a level at which the heat radiated back to space would equate to 288 watts per square meter. Without strictly defined greenhouse gases, and without water vapor, this temperature would amount to –6°C (21.2°F).

The proportion of water in the atmosphere can hardly be disregarded in a comparative scenario, though, because it depends on the temperature that drives the evaporation of the ocean's waters. Moreover, account has to be taken of the fact that lower temperatures lead to more condensation of water vapor in the atmosphere, which diminishes the amount of solar radiation reaching the surface. In the absence of strictly defined greenhouse gases, but with water in the atmosphere, the average temperature would be –18°C (– 0.4°F).[9] Not only would our planet be as cold as Siberia; the cloud cover would let hardly a sunbeam through.

The nearly 32°C (58°F) temperature increase from –18°C to +13.5°C that makes our planet inhabitable comes mainly from the fact that carbon dioxide and the other narrowly defined greenhouse gases trap some of the radiated heat. Figure 1.1 schematizes this relationship. The higher temperature leads to increased evaporation from the oceans, and with the higher water vapor content in the atmosphere a further greenhouse effect comes into play. With the increased warmth, there is less cloud formation, which in turn accelerates warming. Though the clouds block some of the radiated heat, the fact that they reflect sunlight exerts a larger effect. All in all, during the pre-industrial period, with the greenhouse gases present then, a temperature of +13.5°C (56.3°F) was necessary to radiate back into space exactly the 343 watts per square meter that the planet received from the sun.

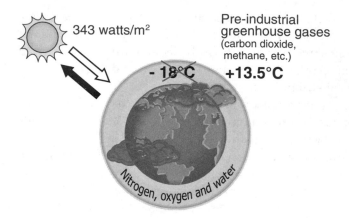

Figure 1.1
Warming the Earth.

Why it all comes down to carbon dioxide

Greenhouse gases aren't all alike. Each has its peculiarities, and these must be understood in order to ascertain their meaning for our climate.

Water vapor is the most important greenhouse gas. Though its contribution per molecule to the greenhouse effect is equal to only 4 percent of the contribution of carbon dioxide, there is so much water vapor in the atmosphere that it accounts for about 65 percent of the total effect. With about 2.5 percent per volume (that is, 25,000 ppm), it is by far the most abundant climate-relevant gas in the atmosphere.[10]

But given that water vapor concentration in the atmosphere is endogenously determined by the Earth's temperature, water vapor usually isn't included among the greenhouse gases. This gives rise to the distinction between strictly and broadly defined greenhouse gases mentioned above. Though water vapor has an enormously important feedback or self-reinforcement effect in greenhouse mechanics, it isn't an autonomous determining factor that can be changed by the hand of man, other than through temperature itself.

This is important in view of the common assertion that the influence of carbon dioxide is irrelevant relative to the overwhelming importance of water vapor. Instead of focusing on CO_2, the argument goes, we should pay attention to the fact that enormous amounts of water vapor are expelled into the atmosphere from the cooling towers of power plants and through the burning of hydrocarbons such as coal, natural gas, and oil. Furthermore, attention ought to be paid to the fact that a hydrogen-based economy, which would release significant amounts of water vapor into the atmosphere, would warm the planet even more. These assertions ignore the fact that the proportion of water vapor in the atmosphere regulates itself continuously through the weather. Water vapor is released from the oceans, condenses, rains down again within 8 to 10 days, and is brought back to the sea by the rivers.[11] How much of this constantly circulating water vapor remains in the atmosphere and contributes to the greenhouse effect depends on the temperature of the air. The warmer the air, the more water vapor it can store. You can see it any morning: dew evaporates as the air temperature increases. Any amount of additional water that human activities pump into the air will quickly rain down again and thus can't contribute to the greenhouse effect.

Carbon dioxide, in contrast, plays a central role in the influence that human activities are exerting on the climate. Though it ranks second to water vapor as a greenhouse gas and accounts for about 60 percent of the third not accounted for by water vapor (see table 1.1), CO_2 is incomparably more important than water vapor in explaining climate change, because its content in the atmosphere isn't determined solely by natural processes but rather keeps increasing as a result of human activities.

Carbon dioxide readily binds with water vapor, forming carbonic acid, and gets washed into the oceans when it rains. Waves then release it back into the atmosphere, in a fashion similar to the bubbling away of carbon dioxide when you shake a soda bottle. However, this exchange process has little similarity to the water cycle, as the amount of CO_2 that the atmosphere can absorb isn't limited by natural forces; it can be increased almost indefinitely by human activity. What is limited is the capacity of the oceans' upper layers to absorb it. As the concentration of CO_2 increases in these layers, which are responsible for the exchange between water and air, a larger amount of CO_2 will be released by the waves. Thus, only a limited amount of the CO_2 released by human activities can be absorbed by the seas; the rest accumulates in the atmosphere and in biomass.

As was noted above, water vapor gives rise to a feedback effect in climate because higher temperatures lead to more water vapor, which in turn increases the greenhouse effect. Carbon dioxide shows a similar reinforcing pattern. As the temperature of the oceans rises, their capacity to absorb CO_2 decreases. We know this phenomenon from the spraying that happens when we open a warm soda bottle. If external factors bring about a rise in the Earth's temperature, the seas release more carbon dioxide, increasing the concentration of this gas in the atmosphere and thus exacerbating the Earth's warming. We could call this the "fizzing effect." The fizzing effect—that is, the reduced capacity of the seas to absorb CO_2 in the presence of increasing temperatures—is the most important destabilizing factor for our climate.

Another destabilizing factor would be the thawing of the permafrost regions, most of which are in Siberia and Canada. Should this happen, a decay processes in those tundra regions would take place, releasing carbon dioxide and methane and thus accelerating the greenhouse effect. Fortunately, these destabilizing factors aren't strong enough to bring the planet's temperature to a tipping point. Since the stabilizing effect of higher amounts of infrared wavelengths being radiated back to space after an increase in temperature is significantly stronger, the resulting greenhouse effect would bring about a considerable increase in the planet's temperature but not an uncontrolled, runaway one.

Carbon dioxide is absorbed not only by the oceans but also by plants. If the atmosphere contains higher proportions of CO_2, some plants grow faster, as a nutrient relevant to their growth would be more abundantly available. On the other hand, the subsequent decay of those plants would release higher amounts of carbon dioxide as well. Still, more carbon dioxide will be absorbed than released, as the biomass stock in terms of plants and animals would increase. This would slow the pace at which temperature increases and thus act as a climate stabilizer. More vegetation can only slow the release of CO_2 into the atmosphere, however; it can't stem it. The vegetation effect is much too weak to prevent temperature increases, not least because high enough temperatures can also cause plants to die off.

Carbon dioxide is much more chemically stable than other greenhouse gases. It doesn't react with other gases in the air, and therefore it doesn't break down. It is being pumped into the atmosphere as a result of the burning of fossil fuels, adding to the stocks already there. This is the principal reason it plays such a significant role in concerns about our climate. Only when the carbon dioxide washed by rain into the oceans enters into a reaction with calcium, building calcium carbonate, and gradually sinks to deeper layers, can the stock of CO_2 in the atmosphere be reduced, but these processes, from our human perspective, are far too slow to pose a solution to climate change. The average time CO_2 emitted today would stay in the atmosphere ranges from 30,000 to 35,000 years.[12]

Other greenhouse gases

Another important greenhouse gas is methane (CH_4), with a concentration of 1.8 parts per million. Methane is a natural gas, most of which leaks out of underground deposits. But it also arises from the natural decay of organic matter. If oxygen is present in the decay process, CO_2 is produced. If decay occurs in the absence of oxygen, methane is produced. This is the case mainly in humus layers, but can also occur through fermentation in the stomachs of ruminants. Methane absorbs much more radiation per unit of weight than CO_2 does. Fortunately, it reacts with oxygen and decays into water and CO_2 in about 15 years on average. It continues to be damaging for our climate, but not as much as it was before decaying. For this reason, the usual practice is to measure its contribution to global warming not in terms of its current absorption of radiation but in terms of the total contribution that one kilogram of methane, in comparison to one kilogram of CO_2, makes to global warming over a certain span of time. Measured this way, methane's greenhouse effect per unit of weight amounts to 72 times that of CO_2 over 20 years, 25 times that of CO_2 over 100 years, and 8 times that of CO_2 over 500 years.[13] Per molecule, its greenhouse effect over 100 years amounts to 9 times that from CO_2. The last figure is important because it helps us understand the consequences of burning methane. Because burning one molecule of methane yields one molecule of CO_2 and two molecules of water, and because excess water is quickly removed from the atmosphere as rain, burning reduces methane's greenhouse effect over 100 years by about a factor of 9. For this reason, we should never allow methane to reach the atmosphere unburned. It is a pity that in drilling for oil the escaping gas is burned rather than put to profitable use, but burning it is better than pumping it into the atmosphere. Farmers must also be congratulated for turning organic waste into gas that they can use for heating, instead of just letting it rot away unattended.

Nitrous oxide (N_2O), with 0.3 ppm, ozone (O_3), with 0.015–0.050 ppm, and the CFCs, with 0.0009 ppm, also have some importance for climate.[14] Nitrous oxide is produced mostly through the use of fertilizers in agriculture. Ozone, which occurs naturally at altitudes of 20–40 kilometers (12–25 miles), forms a barrier against ultraviolet radiation. Ozone is also a major component of summer smog, which occurs at ground level mainly as a result of car exhaust's reacting to sunlight and which is quite unhealthy. Over 100 years, the greenhouse effect of a unit of weight of nitrous oxide amounts to 298 times that of CO_2, and that of a unit of weight of ozone amounts to as much as 2,000 times that of CO_2. But because their concentrations are low, these greenhouse gases account for only 13 percent (N_2O 4 percent; O_3 9 percent) of the anthropogenic (man-made) greenhouse effect.

Chlorofluorocarbons are somewhat more important. The term "chlorofluorocarbons" refers to a group of gases with comparatively complex chemical formulas that cause a great deal of damage in the atmosphere because they destroy the ozone layer that not only contributes to the greenhouse effect but also protects us from ultraviolet radiation from the sun. CFCs are synthetic gases; that is, they aren't found in nature. They were once used in spray cans and in refrigerators, from which they leaked out into the atmosphere. Since 1987, when their production was banned through the Montreal Protocol, the stocks of these gases in the atmosphere have been gradually decreasing, and the ozone holes over the poles are starting to close again. The proportion of one of these gases, CFC-11, reached its peak in 1993 and has been slowly decreasing ever since. No reduction in the concentration of CFC-12 can be detected yet, but since the early 1990s its rate of increase has diminished markedly.[15] The CFCs are important for climate change because relative to their size they produce an enormous greenhouse effect—between 5,000 and 10,000 times as strong per molecule, and up to 11 times as strong per unit of weight, as carbon dioxide's. Over the next 100 years, the CFCs already released into the atmosphere will account for about one-ninth of global warming.

Over 100 years, the current combined greenhouse gases, excluding carbon dioxide and water vapor, will produce a greenhouse effect equivalent to 50–70 ppm of carbon dioxide. That is why at present the "CO_2-equivalent" greenhouse-gas concentration, without water vapor, amounts to approximately 430–450 ppm.[16]

Table 1.1 provides an overview of the current sources of greenhouse gases. The first column gives the volume shares of the various gases in the atmosphere. The second column gives their average permanence in the air. The values range from two months for ozone to 35,000 years for CO_2. The third column gives the greenhouse effect of a kilogram (2.2 pounds) of the respective gas over the next 100 years relative to a

Table 1.1
The greenhouse gases (other than water vapor). Sources: C. D. Schönwiese, *Klimatologie*, second edition (Ulmer, 2003), p. 337; S. Solomon et al., *Climate Change 2007: The Physical Science Basis* (Cambridge University Press, 2007); L. K. Gohar and K. P. Shine, "Equivalent CO_2 and its use in understanding the climate effects of increased greenhouse gas concentrations," *Weather* 62 (2007): 307–311.

Greenhouse gas	Concentration today (ppm)	Average life (years)	Greenhouse potential per unit of weight over 100 years	CO_2 equivalent concentration today (ppm, 100 years)	Percentage of greenhouse effect (100 years)
Carbon dioxide (CO_2)	380	30,000–35,000	1	380	61%
Methane (CH_4)	1.8	15	25	26.3	15%
CFC	0.0009	100	1,810–10,900*	14.3	11%
Ozone (O_3)	0.015–0.05	0.16 (2 months)	<2,000	18.9	9%
Nitrous oxide (N_2O)	0.3	114	298	8.5	4%

*CFC-11, CFC-12, CFC-22

kilogram of CO_2. The fourth column gives the CO_2 equivalent in particles per million for the respective gases according to their current concentration in the atmosphere. This forms the basis for the calculation of their greenhouse effect over a period of 100 years. The last column, which gives the percentage share in the greenhouse effect of each gas, except water vapor, illustrates starkly why such overwhelming significance is attached to carbon dioxide in devising appropriate climate policies.

The human influence

Thanks to the many air samples from times past that nature has left us, we know fairly accurately how much the proportion of carbon dioxide in the atmosphere has changed since the Industrial Revolution. Such air samples are found in air bubbles trapped between rock layers and in the ice of glaciers and the polar caps. The deeper you drill, the older the sample.

The ice cores drilled on the Law Dome in East Antarctica give particularly good measurements.[17] They show that the concentration of CO_2 remained nearly constant at 280 ppm and began to increase sharply after the year 1800, rising to the present-day value of 380 ppm. There is no other explanation for this increase than industrialization. Burning fossil fuels—at first mostly coal, then, starting around the end of the nineteenth century, also oil—has left its traces on our planet. Natural gas, because of its lower consumption and its high content of hydrogen, has thus far played only a minor role. Clearing forests has played a role, however— see chapter 3.

Cement also has been of some importance, as its production releases large amounts of CO_2. When calcium carbonate, the raw material for cement, is heated, it produces lime and CO_2. This process is called calcination. The CO_2 released during calcination comes in addition to that released by the fuels used to fire up the kilns where calcination takes place. This makes cement production very detrimental to our climate. Even under optimum conditions, producing one ton of cement releases 1.4 tons of CO_2. Cement production currently accounts for 4 percent of worldwide anthropogenic CO_2 emissions.[18]

Figure 1.2 shows how industrial CO_2 emissions into the atmosphere have increased. The curve is dramatic. Just since World War II, industrial CO_2 emissions have increased fivefold, and the rate appears to be accelerating.

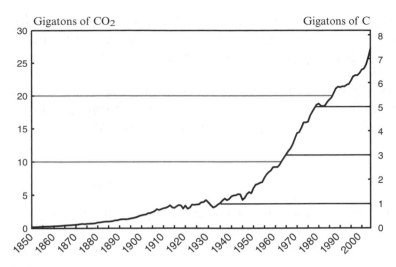

Figure 1.2
Annual global carbon emissions, with CO_2 emissions resulting from changes in land use, such as clear-cutting of forests, not considered. Source: *Climate Analysis Indicators Tool* (CAIT), Version 5.0, World Resources Institute, 2008, converted to gigatons of CO_2 (GtC).

The left-hand scale in figure 1.2 shows gigatons of carbon dioxide; the right-hand scale shows the gigatons of carbon contained in this CO_2. A gigaton is another name for a billion metric tons. In order to simplify comparisons with the carbon stocks still underground and with the carbon contents of fossil fuels, this book will normally base the weight specifications on the carbon content of CO_2. Because CO_2 contains two oxygen atoms attached to each carbon atom, and each oxygen atom is 1.33 times as heavy as a carbon atom, it is easy to establish a proportion between these two weights. You need only multiply the weight of the carbon contained in a specified amount of CO_2 by 3.66 in order to obtain the weight of that amount of CO_2. The chart shows that in 2005 the combustion of fossil fuels and the production of cement released 7.4 gigatons of carbon worldwide, equivalent to 27 gigatons of CO_2.[19]

The curve in figure 1.2 shows the yearly flow of carbon dioxide emissions. The area underneath the curve is the total stock of CO_2 emitted. For the greenhouse effect, what is important, of course, is not the stock emitted, let alone the annual flow of emissions, but the emitted stock that has not been absorbed by the oceans or by biomass. With forest clear-cutting included, humans have increased the stock of carbon in the atmosphere from about 600 to 800 gigatons of carbon since the Industrial Revolution, which corresponds to an increase in CO_2 density from 280 to 380 ppm.

Whether public policies can induce people to extract and release to the air less fossil carbon so as to slow down global warming, and if so how they can do so, is the main theme of this book, especially of chapters 4 and 5.

One more degree already

The big question is this: How is the anthropogenic increase in greenhouse-gas concentration affecting the temperature of our planet's surface? In view of the large temperature fluctuations between hot periods and ice ages that have always occurred, it is legitimate to ponder to what extent mankind has provoked the current climate change. Research has shown, for example, that in the second half of the fifteenth century a small ice age lowered temperatures in Europe by about 0.3°C (0.5°F). Three hundred years earlier, temperatures were 0.2°C (0.4°F) warmer than usual. About 18,000 years ago, when the last ice age came to an end, the world was 5.5°C (10°F) colder than it is now.[20]

A further difficulty in isolating the effect of human activity is that the temperature reacts very slowly to a change in heat radiation from the Earth. The new stable equilibrium resulting from the change in the greenhouse-gas concentration mentioned above will be reached very quickly from a geological point of view, but from our human perspective it appears to take much longer. It takes many decades until the warming process of air, rocks, and bodies of water has been completed. Whereas the temperature of the air in the layers in which airplanes fly (the troposphere) can change within a few days, in the stratosphere it can take much longer. Much slower still is the change in temperature at surface level, as the oceans, which react very sluggishly to climate change, exert a dominant influence there. Estimates say that the greenhouse gases already emitted since the pre-industrial period will raise the Earth's temperature by about 0.5°C (0.9°F) over today's average by the end of this century, even if the greenhouse-gas concentration were to remain stable from now on.[21]

However, despite this sluggishness, the Earth has already become noticeably warmer. This is evident from many indicators, among them

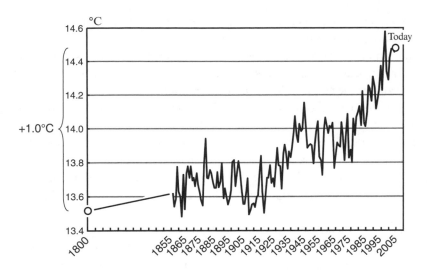

Figure 1.3
Global average temperatures measured by thermometers. Note: The global average temperature is calculated on the basis of measurements from 3,000 weather stations. The dot near the bottom of the left scale shows the pre-industrial temperature level in the year 1800, which according to Otto-Bliesner et al. amounted to 13.52°C (56.34°F). The most recent value for the current global average temperature, according to Jones et al., amounts to 14.48°C (58.06°F); this is depicted by the dot near the right-hand scale. Sources: P. D. Jones et al., "Global and hemispheric temperature anomalies—land and marine instrumental records," in *Trends: A Compendium of Data on Global Change*, Carbon Dioxide Information Analysis Center, Oak Ridge National Laboratory, 2006; B. L. Otto-Bliesner et al., "Last glacial maximum and holocene climate in CCSM3," *Journal of Climate* 19 (2006): 2526–2544.

direct temperature measurements made since the invention of the thermometer. Though there have been many fluctuations, the planet's average temperature seems to have already increased by about 0.7°C or 0.8°C since 1855, and has now reached 14.48°C (58.06°F).[22] Complex measurements using other indicators such as climate models and temperature anomalies even show an increase of nearly exactly 1°C (1.8°F) since the year 1800, when the average temperature was 13.52°C (56.24°F).[23] Figure 1.3 illustrates this.

The temperature curve in figure 1.3 should be interpreted with some caution. Measurement methods may have changed over time, and it isn't entirely clear whether the data have been properly adjusted for the effect of urbanization on the air surrounding measurement stations. At present, however, the curve depicted in the figure reflects the best available direct data on air temperatures.[24]

The temperature didn't increase equally everywhere; it rose more in inland areas than on those close to the coast, and more with increasing distance from the equator. In Germany, for example, a station in Potsdam shows an increase from 1890 to today from 8.3°C to 9.9°C (46.9 to 49.8°F), i.e., a 1.6°C rise, much higher than the global average.[25]

It is remarkable that the warmest ten years since the invention of the thermometer all occurred in the last eleven years (before this book first appeared in German). They were, in decreasing order of global average, the years 1998, 2005, 2003, 2002, 2004, 2006, 2007, 2001, 1997, and 1999.

Figure 1.3 shows that the temperature increase took place in two surges: one from 1900 to around 1945 and one since 1975. Over the period 1945–1975, practically no temperature increase was registered. This can probably be attributed to the high emissions of sulfur dioxide during the fast-paced economic development that took place after World War II. Sulfur dioxide is emitted when coal and oil are burned, and then is transformed into sulfate particles that block sunlight, leading to cooler temperatures on the Earth's surface. After measures to reduce air pollution and sulfur dioxide emissions were introduced globally, starting in the 1970s, this effect disappeared and the greenhouse effect took over again.[26] This interruption in global warming explains why the greenhouse effect only recently became of interest to both scientists and public opinion, and why it still was irrelevant during the oil crises of 1974 and 1982.

Whether there really is any anthropogenic global warming has been a subject of heated debate in recent years. Some skeptics, among them Scafetta and West,[27] argued that the increase in the temperature was due to a strong increase in the sun's radiation since about 1900. However, this argument was refuted by Benestad and Schmidt, who showed that only 8 percent of global warming in the twentieth century can be explained by that effect.[28] Lockwood and Fröhlich even showed that in the last quarter of the twentieth century, when the temperature was rising particularly rapidly (see figure 1.3), all changes in solar activity that in principle could have affected the temperature went in the "wrong" direction, slowing rather than accelerating global warming.[29]

Other skeptics claimed that the temperature measurement indicating global warming since the Industrial Revolution was flawed. Their criticism gave rise to the "hockey stick" controversy. A long-term data set provided by Mann, Bradley, and Hughes[30] had gained much popularity after being featured prominently in an IPCC report. The authors had reconstructed a temperature curve from proxy indicators, such as the widths of tree rings, the calcification rates of coral, and the composition of sediments, that showed that the temperature hadn't changed much during the last millennium except in the last 100 years. The curve representing the data looks like a hockey stick lying horizontally with its blade pointing up.

McIntyre and McKitrick[31] argued that the data set of Mann, Bradley, and Hughes was useless because the method for creating a temperature indicator from the proxy data was mistaken. They demonstrated that with the method that had been used to transform the proxy into temperatures even randomly chosen numbers would have produced a "hockey stick."[32]

The issue was discussed and re-investigated by a great many authors. Their findings are summarized in a report published by the National Research Council in 2006.[33] According to that report, nearly all other authors who delved into the issue found the global warming effect in proxy data, and the overwhelming majority confirmed the magnitude of the temperature rises of Mann et al. for the last 400 years, whereas there was more ambiguity for more remote periods of time. The data screened referred to, among other things, bore-hole temperatures, glacier lengths, tree rings, and various composite proxy indicators.

Meanwhile, the criticized authors invited others to join a bigger research program that would reconsider the issue. After correcting their previous mistakes, they presented a revised data set that resulted in essentially the same kind of "hockey stick" curve as before. This new data set was again criticized by McIntyre and McKitrick, and again the criticism was refuted by the original authors. There is an ongoing debate that, at this writing, has not yet come to a conclusion.[34]

Whatever its final outcome, this debate concerns only one of many data sets that have been collected and screened. The data sets leave little doubt about the global warming effect. One such data set is the direct measurement of temperature by thermometers, as in figure 1.3. The curve shown there isn't subject to the criticism of McIntyre and McKitrick, and doesn't contain proxy data.

The past 800,000 years

Truly fascinating data were obtained from ice-core drilling done in Antarctica by an international team of researchers working under the auspices of the European Union's Project EPICA. The researchers managed to bore into the Dome C ice mountain to a depth of 3,270 meters (10,726 feet), reaching 800,000 years into the past and thus amply surpassing the previous record of 650,000 years.[35] They obtained data on the CO_2 concentration in the atmosphere and on air temperatures.

The CO_2-concentration data came from air bubbles trapped in the Antarctic ice, which consisted of compressed snow that had accumulated without ever melting.

How the temperature data were obtained is less straightforward. After all, ice is equally cold everywhere. But temperature data can be reconstructed by means of the isotope method, an ingenious method that has fundamental significance in climate research. Water consists of hydrogen and oxygen atoms, which aren't homogeneous; they can vary according to the number of neutrons contained in their nuclei. A variation resulting from different numbers of neutrons in the nucleus is called an *isotope*. Water containing oxygen 16 and water containing oxygen 18 evaporate at different speeds. Oxygen 16 makes for lighter water, oxygen 18 for heavier water. Because lighter water evaporates faster than heavier water, the ratios of the two isotopes in Antarctic ice cores provide precise indications of the temperatures that prevailed in previous periods. These ratios reveal the temperature of seawater that subsequently evaporated, was transported by the wind over Antarctica, precipitated there as snow, and eventually became compressed into Antarctic ice. The great advantage of the isotope method is that it can be carried out using the same ice cores used to determine the CO_2 content of the air bubbles trapped in them. This makes it possible to derive data on CO_2 content and temperature from the same sample. (See figure 1.4.)

At present, the Earth's average temperature on the surface is 14.5°C (58°F), as mentioned above. This is the highest average temperature not only in the past few years but in the past 100,000 years. It was last higher during the Eemian Interglacial, a warm period that began 128,000 years ago and lasted 11,000 years. That was the warmest period in the last 800,000 years, during which humankind evolved from *Homo erectus* to *Homo sapiens*. During the Eemian Interglacial, the Earth's average surface temperature exceeded the average of the past 800,000 years, which was about 11°C, by 4°C—that is, it reached about 15°C (59°F). With our 14.5°C, we are close to this. Obviously we are living through one of the warmest periods in human history.

Figure 1.4 shows a temperature curve covering the last 800,000 years. The time axis requires a bit of mental adjustment, as each space between

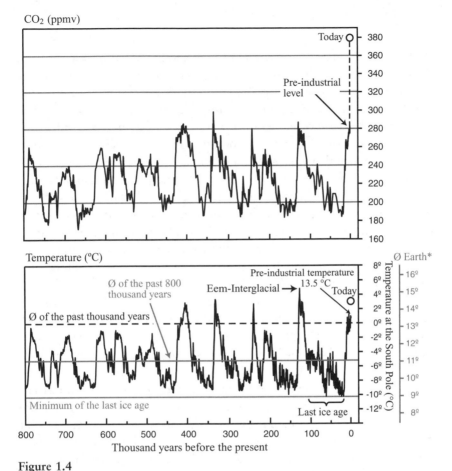

CO₂ (ppmv)

Temperature (°C)

Figure 1.4
Ice-core drillings. The temperature scale on the right side of the upper panel shows the deviation of Antarctic temperature from its average of the past 1,000 years; that on the right side of the lower panel shows the corresponding average temperature. This scale was based on other data that were adjusted according to the following information: (1) The last value in the temperature curve shows the Antarctic temperature in the year 1912. This temperature lies exactly 0.88°C above the average of the past 1,000 years. (2) The lowest Antarctic temperature during the last ice age (about 18,000 years ago) was 10.2°C (1.56°F) below the mean of the last 1,000 years. (3) In 1912, the average temperature was 13.6°C (56.5°F), 0.1°C above the "pre-industrial" temperature in the year 1800; see figure 1.3. (4) According to Otto-Bliesner et al., the average temperature during the coldest period of the last ice age was 8.99°C (48,18°F). Sources: D. Lüthi et al., "High-resolution carbon dioxide concentration record 650,000–800,000 years before present," *Nature* 453 (2008): 379–382; J. Jouzel et al., "Orbital and millennial Antarctic climate variability over the last 800,000 years," *Science* 317 (2007): 793–796; B. L. Otto-Bliesner et al., "Last glacial maximum and Holocene climate in CCSM3," *Journal of Climate* 19 (2006); 2526–2544; own calculations.

bars represents 50,000 years. The numbers are counted backward from the present. The "current" average CO_2 level, spanning the last 500 years, is shown on the right side of the CO_2 curve. It amounts to 280 ppm, which, as was mentioned above, corresponds to its pre-industrial value.

Figure 1.4 has two temperature scales, both on the right side. One shows deviations in Antarctic temperature from the average of the past 1,000 years; the other translates these temperatures into a global average temperature. The zero value on the left-hand scale corresponds to a 13.3°C (55.9°F) global average on the right-hand scale. According to figure 1.3, this average was slightly lower than the one prevailing during pre-industrial times, i.e., 13.5°C (56.3°F). The right-hand temperature scale is somewhat more stretched than the left-hand one, because the average temperature over most parts of the planet doesn't show such strong swings as the temperature in Antarctica.

The last ice age can be clearly made out in the period from 115,000 to 10,000 years ago; it reached its coldest period 18,000 years ago, with an average of just 9°C (48°F). The Eemian Interglacial also can be seen clearly. It occurred around 125,000 years ago, with an average temperature of 15.3°C (59.5°F).

It is striking how much the Earth's temperature has fluctuated. The fluctuations can be attributed to disturbances that affect the amounts of energy received and radiated. One of the disturbances is the regular displacement of the Earth's axis. Like a top, the Earth wobbles as it rotates more slowly. The Earth's axis wobbles at a rate of about one spin every 26,000 years. This wobbling leads to a change in the amount of radiation received by the planet's darker and lighter areas, changing the amount of heat they absorb. Meteorite impacts and volcanic eruptions can also influence our climate: the dust they release into the atmosphere reduces the amount of sunlight reaching the surface, while the carbon dioxide they also release works in the other direction, increasing the greenhouse effect. In addition, solar radiation has itself experienced large variations over time, as shown by the sunspot phenomenon. All these factors have combined to create the fluctuations in our climate depicted in figure 1.4.

Today's values, those measured directly from the atmosphere, are also shown in figure 1.4. They amount to 380 ppm of carbon dioxide and an average global temperature of 14.5°C (58.1°F). These values, when set against the Earth's geological ages, illustrate the uniqueness of our current situation. There was never, over the 800,000 years shown, as much carbon dioxide in the atmosphere as there is now,[36] and the temperature now also hovers in the upper ranges, exceeded only occasionally during the interglacial warm periods.

Correlation and causality: A solvable puzzle

It is striking how closely correlated the CO_2 curve and the temperature curve are. On the face of it, one could take this correlation as proof of the greenhouse effect. It looks as if the changes in the atmosphere's CO_2 content have influenced the temperature. Closer examination, however, shows that this can't be correct, as most of the temperature extremes occurred a bit earlier than the corresponding peaks in CO_2 content. The data show that the Earth's temperature changes occurred, on average, about 800 years before the changes in the atmosphere's CO_2 content. This eliminates the possibility that the greenhouse effect was the major force behind the correlation observed.

The real reasons for the correlation are found in the fizzing effect, in the permafrost effect, and in biological processes. These effects were mentioned previously in relation to the self-energizing processes of the greenhouse effect. When external influences that increase incoming solar radiation lead to an increase in Earth temperatures, the oceans' capability for storing CO_2 decreases, their waves instead transferring this gas into the atmosphere. The permafrost areas, in turn, begin to thaw and release CO_2 through the decay of organic matter, either directly or through the production of methane, which quickly decays through oxidation into CO_2. With higher temperatures, the carbon stored in biomass will also be reduced, as deserts will expand. The opposite occurs when temperatures decrease. In that case, the oceans will again absorb more CO_2 and, up to a certain point, more plants will grow, their photosynthesis capturing CO_2 from the atmosphere and storing it as biomass. All these effects explain why temperature oscillations bring about a corresponding fluctuation in the atmosphere's CO_2 concentration.

Some skeptics have used these findings to cast doubt on the assertions about the effects of greenhouse gases on our climate. It is claimed that climate researchers have manifestly misinterpreted the correlation observed by attributing temperature fluctuations to the variations in the atmosphere's CO_2 content. Thus, skeptics say, the entire climate discussion that has caused such an uproar around the world is based on a fallacy, and we should therefore desist from introducing measures to reduce industrial CO_2 emissions.

These arguments are hollow. No serious climate researcher has ever asserted that the correlation between CO_2 content in the atmosphere and temperature was due to temperature-independent disturbances in CO_2 content in the atmosphere. In fact, climate researchers have arrived at their conclusions essentially through theoretical models and refined statistical methods that stand above the allegation that they are based on a mere interpretation of the correlation observed.

As was explained above, the central element in the theoretical explanation is the absorption of infrared back-radiation by the greenhouse gases. This absorption is a physical effect firmly established in theory and confirmed by many experiments. Direct proof of this absorption has been provided in recent years by spectral measurements from satellites. Since the satellites are outside the atmosphere, they make it possible to measure the back-radiation behind the "atmospheric filter." The measurements show that the spectral frequencies CO_2 is known to absorb from theoretical considerations and experiments are indeed being absorbed, thus warming the atmosphere.[37]

A particularly interesting result was published in 2001 by Harries, Brindley, Sagoo, and Bantges,[38] who compared spectral measurements made in 1970 by a NASA satellite against spectral measurements made in 1997 by a Japanese satellite. After showing that the two data sets were comparable, Harries et al. found that the relevant infrared back radiation filtered out by the greenhouse gases had been substantially reduced over the measurement period. Thus, the temperature increase over this period can indeed be largely attributed to the greenhouse effect.

It is true that the correlations shown in figure 1.4 are due predominantly to the permafrost and fizzing effects rather than to the greenhouse effect. But the reason is simply that variations in solar radiation are bigger and more frequent than exogenous variations in greenhouse-gas concentration in the atmosphere, for which, except for the human influence, little other than volcanic eruptions would come into consideration. The predominance therefore doesn't imply that there is no greenhouse effect. Usually a car comes to rest because the driver has stepped on the brake pedal. More seldom, one comes to rest because it has run into an obstacle. The fact that the former cause empirically occurs more often than the latter doesn't mean that the latter cause is irrelevant and not worth trying to avoid.[39]

The changes in CO_2 content in the atmosphere caused by temperature variations have not caused the climate fluctuations over geological periods, but they have amplified them. The fact that these fluctuations have attained the magnitudes depicted in figure 1.4 is also due to an accelerator or feedback effect stemming from the greenhouse gases that the rising temperatures caused the ocean water to expel.[40] Even though exogenous variations in CO_2 were rare before industrial times, CO_2 has always played an important part in making the temperature variations as large as they were. When the oceans were being warmed by increased radiation from the sun and were releasing more CO_2, the additional CO_2 then warmed the planet even more. This is reason enough to be afraid of the exogenous variations brought about by industrialization.

Unfortunately, as was mentioned above, we are currently living through a warm period. If we were in the midst of an ice age, the additional global warming resulting from industrial greenhouse gases would be quite welcome. It would counteract the geological cycle and exert a stabilizing effect. The reality is, however, exactly the opposite. During pre-industrial times the global temperature was already above its long-term average. Now man-made effects are exacerbating the increase, bringing it to levels resembling the peaks of the past million years.

We are burning carbon stocks that were essentially formed during the Carboniferous period, from 280 million to 340 million years ago, from vast forests. The burial of large forest areas as a result of tectonic movements led to the formation of bogs in which new plants grew, died, and gradually sank ever deeper. The resulting coal, oil, and natural gas were removed from the biological cycle until man began pumping them back into that cycle once again.

The stocks of fossil fuels played no part in the fluctuations of CO_2 content in the atmosphere shown in figure 1.4. They lay so deep that no oxygen could reach them; thus, they were not able to burn or otherwise oxidize and thus release carbon dioxide. These fluctuations were essentially results of the displacement of a given amount of carbon between the oceans, biomass, and the atmosphere brought about by the fizzing water effect and by biological processes. Only volcanic eruptions on the planet's surface increased the amount of carbon in circulation, but that effect was relatively marginal. Volcanic emissions account for only one-tenth of a gigaton of carbon per year, equivalent to only 1.25 percent of the yearly anthropogenic emissions.[41]

The climate-change problem arises from the fact that mankind has increased the amount of carbon cycling through the atmosphere, the oceans, and biomass by adding fossil carbon that formed during the Carboniferous and had lain undisturbed, not taking part in nature's carbon cycle, for millions of years. Over the next 500 years—a vanishingly short portion of the time span represented in figure 1.4—we will tap, and perhaps exhaust, a reservoir that took about 120,000 times as long to form. This will lead to a break in the trend of the temperature curve and will cause a lasting increase in the average measured over ice ages and interglacial periods, regardless of variations caused by disturbances in solar radiation that bring about fluctuations between warm and cold periods. Of course it is possible that, as was the case during the Carboniferous, tectonic changes in the Earth's crust will again remove carbon permanently from the carbon cycle. That, however, should not be expected in, say, the next 800,000 years. Humans will have disappeared from this planet before the new carbon is removed from the cycle and again stored in the crust.

On to the North Pole

The first consequences of global warming are visible in many places. Photographs taken during the first half of the twentieth century show clearly that the glaciers in the Alps are retreating. The area of the Watzmann glacier, for instance, shrank by 64 percent from 1897 to 2006, that of the Northern Schneeferner glacier by 70 percent from 1892 to 2006, and that of the Southern Schneeferner glacier by 90 percent from 1892 to 1999.[42] The Arctic is another case in point. In the years 1996–2006, the area of the ice sheet over the North Pole shrank by 1.5 million square kilometers (580,000 square miles), 23 percent of its total area. The shrinkage was so extreme that during the summer of 2007 the Northwest Passage between Alaska and Labrador was ice-free for the first time, prompting Russia to quickly proclaim its sovereignty over the Arctic.

Our planet warms more rapidly over the North Pole than over the South Pole because it has more land surface and less ocean in its northern portion, even though the South Pole lies over a continent and the North Pole over an ocean. The South Pole has been so cold, and will remain so cold for the foreseeable future, that its ice can't be expected to melt during this century.[43]

So far, because only the North Pole's ice is melting, the sea has risen very little from its level during the pre-industrial period. Ice that melts in the ocean can't raise the water level. If an iceberg melts, the melted water exactly fills the space below the surface that the iceberg displaced before. One cause of the rise in sea level is the melting of glaciers in the Northern Hemisphere. The most significant contributor in this regard would be Greenland's ice cap. Greenland was settled during a warmer period, around the year 1,000, by a Viking known as Eric the Red, but then it turned too cold for further settlement. Today, once again, it is literally flourishing. Even orchids are blooming there now. The forecasts are that an increase of the global average temperature of about 2 to 3 Celsius degrees relative to pre-industrial times will cause the Greenland ice cap to begin to shrink.[44] If all ice disappeared there (something that would take hundreds of years), the sea level would rise by about 6 meters. Another cause of sea-level rise is that water expands as it gets warmer. The two effects combined have caused only a 20-centimeter rise in sea levels until now,[45] but more is to come.

How warm will it get?

How much will the temperature on the surface of our planet rise, and what consequences will that have for life?

We should not expect the worst. Life on our planet will not be wiped out by the greenhouse effect. The relevant models can allay our fears. Earth's physical properties make a runaway process like that on Venus impossible. Such a runaway process would be imaginable if the human-caused increase in the atmosphere's carbon dioxide content were to escalate to a self-energizing reaction that would lead to ever more heat, ever more water vapor, and a release of the carbon dioxide contained in the oceans until the planet literally began to boil.[46] One reason that can't happen is that Earth receives only half as much solar radiation as Venus.

Nevertheless, tipping events could lead to an acceleration of global warming even if the temperature initially rises only a little.[47] If the Greenland ice sheet and the ice on Antarctica no longer cover dark-hued land, more sunlight will be absorbed, and warming will accelerate. A similar effect is operating in the Arctic. Though the melting of the Arctic ice itself doesn't raise the sea level, open-sea water absorbs more solar radiation, which accelerates global warming. Moreover, an increase in temperatures can destroy the complex equilibrium of tree physiology, fire, and rainfall in the boreal forests (coniferous forests in the Russian taiga and northern Canada), in the Amazon rainforest, or in the West African monsoon regions, killing the trees and setting free the carbon that had been captured in them. Melting of permafrost regions results in the release of huge amounts of methane and CO_2, which further accelerates global warming.

What *could* happen is bad enough, as was pointed out by a commission appointed by the British Government and led by Nicholas Stern, a former chief economist at the World Bank.[48] The Stern Commission's report, published in 2007, received widespread attention and has provided a strong impetus to public debate of global warming in recent years.

The Stern Commission examined alternative scenarios for the further evolution of the global climate. In the most likely alternative in their calculations—the business-as-usual (BAU) case—the commission concluded that the carbon dioxide concentration in the atmosphere will have risen from 280 ppm around the year 1800 through today's 380 ppm to 560 ppm by 2050. In the worst-case scenario, this concentration could come to pass as early as around the year 2035.[49]

The rise in temperature associated with this increase in carbon dioxide amounts to 3°C (5.4°F) over the pre-industrial average, i.e., 2°C over today's level. This would be enough to make Greenland's ice cap begin to melt. The temperature on the surface would rise from its pre-industrial average of 13.5°C (56.3°F; in 1800), through today's average of 14.5°C (58.1°F), to 16.5°C (61.7°F). This would be a substantial acceleration of the pace prevailing in the past 150 years. As a comparison with figure 1.4 shows, it would become the highest temperature in 800,000 years. The 15.3°C (59.5°F) record of the past 800,000 years may be broken as early as 2030.

If humanity does nothing, the atmospheric CO_2 content will continue rising unabated. How far it will go is debatable, as no one can predict with certainty how rapidly the world's economy will grow and how the owners of fossil-fuel resources will react to the various incentives they face. Even the best economist with the most sophisticated models can only calculate scenarios on the basis of a set of assumptions that themselves are not predictions but merely plausibility considerations. The Stern Commission investigated a range of very different trend extrapolations published in the literature, in particular those of the IPCC. In what they considered the most plausible scenario for the business-as-usual case, the CO_2 concentration would rise to 900 ppm by 2100, which equates to a temperature of 18.6°C (65.5°F), 5.1°C (9.2°F) above the pre-industrial level.[50]

The Stern Commission's business-as-usual scenario corresponds roughly to the A1FI scenario of the 2001 IPCC Report. According to the latter, the world's population would increase from 6.5 billion to only 7.1 billion by 2100, but global GDP would soar from 48.5 trillion US dollars to about $525 trillion at today's prices. Yearly global carbon dioxide emissions would increase fourfold from their 1990 levels to 30.3 gigatons of carbon.[51] The A1FI scenario is one of a whole family of scenarios investigated by the IPCC. All of them assume that globalization progresses rapidly, leading to fast economic growth and a rapid regional convergence of living conditions. According to the A1FI scenario, the world's population will reach its maximum around the middle of the century, and the developing countries will progress so quickly that their per-capita income will reach two-thirds of that of developed countries. It also assumes that energy will continue to be obtained from the intensive burning of fossil fuels.

A2, an alternative scenario, assumes that the world's regions will not converge so quickly. Developing countries' per-capita income will not rise above one-fourth of that of developed countries until 2100, because those countries will fail to get their population growth under control. The world's population will thus increase to 15 billion by 2100, with GDP rising to only $250 trillion at today's prices. Figure 1.5 illustrates these two projections.

It is to be hoped that neither scenario will prove true, and that humanity will manage to curb CO_2 emissions in time. These scenarios, however, are realistic trend extrapolations of the case where we do nothing and continue business as usual. They aren't even the most pessimistic scenarios. Because we can forecast self-energizing effects in climate only up to a point, things could be significantly worse. The range of dispersion in figure 1.5 shows how far deviations in either direction could go. The most optimistic scenarios project a 3°C (5.4°F) increase in global average temperature over pre-industrial levels; the most pessimistic ones project 6°C (10.8°F).

Lately, reports suggesting that the more pessimistic scenarios are becoming more likely seem to be proliferating. In November 2008, the head of the International Energy Agency (IEA), a research outfit supported by the Organisation for Economic Cooperation and Development,

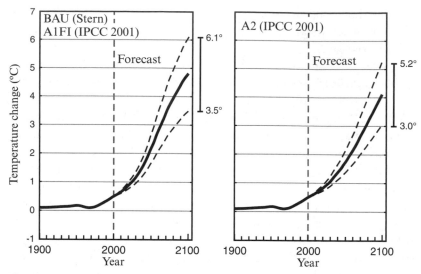

Figure 1.5
Temperature increase forecasts in the absence of a climate policy. The A1F1 scenario of the IPCC TAR Report 2001 corresponds roughly to the Stern Commission's business-as-usual scenario. Dashed lines depict how the projections vary according to changes in assumptions. Sources: J. T. Houghton, Y. Ding, D. J. Griggs, M. Noguer, P. J. van der Linden, X. Dai, K. Maskell, and C. A. Johnson, *Climate Change 2001: The Scientific Basis.* (Cambridge University Press, 2001), p. 554; calculations by H.-W. Sinn.

located in Paris, and employing 190, claimed that the Earth would warm by 6°C by the year 2100 over the pre-industrial average (i.e., 5°C over today's average) if we were to fail to adopt radical measures forthwith.[52] This assertion fits with the Stern Commission's BAU scenario and with the IPCC's A1FI scenario. Nicholas Stern himself has, in the meantime, made even more alarming statements. He calls for carbon dioxide emissions to be halved by 2050 relative to the 1990 levels.[53]

Those who think a 5–6°C temperature increase is not much should bear in mind that this is the amount by which the average temperature has risen on our planet since the peak of the last ice age, 18,000 years ago. What took 18,000 years would now come to pass in only 100.

On top of that, the rise in temperature would not be spread equally across the Earth. The oceans would become 4°C (7.2°F) warmer, Western Europe 6°C (10.8°F) warmer, and northern Finland and Siberia no less than 8°C (14.4°F) warmer. The average temperature would reach 19–20°C (66–68°F), thus exceeding by 4–5°C the maximum attained in the past 800,000 years. That would change how humans live and how they interact. With such temperatures, humanity would enter, in the words of the Stern Commission, "unknown territory."

What is so bad about it?

Not everyone shares these concerns. What would be so bad about the Earth's becoming a bit warmer? What is the big deal about a couple more degrees by 2035 and four or five more by 2100? Isn't it too cold in most of the Northern Hemisphere anyway? Those who find it a bit chilly sitting on their terrace on a summer evening would not mind a couple of degrees more. And, after all, do we not vacation in the sunnier, warmer spots? In the tropics, the air is about 17°C (31°F) warmer than in northern Europe, and even in northern Italy it is 5.5°C (10°F) warmer than in, say, Germany.[54] If the Italian weather were to be transferred to Germany, Germans would be able to save transportation costs on their summer jaunts. It can't be all that bad.

And then, think of Siberia. The largest contiguous land mass stretches from Norway to northern China, but a significant portion of it is too cold for agriculture. The permafrost regions of the former Soviet Union cover an area of 11 million square kilometers, one-tenth more than the land area of the United States. Wouldn't it be a good thing if those places could be a bit warmer, allowing more people to live there? Furthermore, if Russia's Arctic Sea were open to shipping thanks to the polar ice retreating, a brisk sea trade could develop, linking newly flourishing coastal towns with each other. True, Sicily might wither away, but how big is Sicily in comparison with Siberia?

Such views, however, betray superficial knowledge. Technical literature has substantiated the following effects:

• Savannahs and deserts would expand. The subtropical regions, home to many people today, would be assailed by droughts. Droughts affect the entire Mediterranean region even now, but they would extend to Western Africa, Mexico, California, and Australia and make life difficult there.[55] Heat waves like the one in 2003 that caused the death of 35,000 people, 15,000 of them in Paris,[56] would become more common, as would brush and forest fires, which even today regularly sweep through southern Europe, California, and Australia.

- Sea-level rise would continue, exacerbated by the melting of continental ice masses (particularly the Greenland icecap) and by the expansion of sea water caused by higher temperatures.[57] Estimates point to a rise of 1 meter (3.28 feet) from pre-industrial levels by the end of this century. That would cause problems not only in Bangladesh but also in many other coastal areas. The Netherlands would be the most affected country in Europe.[58] The Dutch would surely have no problem raising their dikes another meter, but how they would dump the flow of the Rhine into the North Sea is a matter of heated debate even now.

- Because higher average temperatures mean higher temperature differences between regions, between land and sea, and between air layers, air movements would become much stronger. Hurricanes and typhoons would increase in number and in force.[59] The southern United States and Japan and other Asian island states would be particularly affected.[60]

- The shifting of the habitable regions would lead to migrations akin to those that occurred during the ice ages and the interglacial periods.[61] Large migration flows from south to north could be expected in the Northern Hemisphere. This would not happen without armed hostilities, ethnic conflicts, civil wars, and great social penury. The world would have to find a new settlement structure.[62] This is certainly the greatest potential peril presented by the greenhouse effect.

- Changes in weather patterns associated with climate change probably would affect human health, in particular owing to changes in the distribution of infectious disease vectors and increased risk of respiratory and skin diseases.[63]

There is also the fear that the Gulf Stream might stop flowing if more fresh water were to reach the oceans. That would have dramatic consequences for Western Europe, which, contrary to the general trend, would develop into a cold zone. Europe north of the Alps would be barely inhabitable, with climatic conditions similar to those prevailing in northern Canada. This concern is, however, not shared by most of the literature. At most, a weakening of the Gulf Stream is expected, which would not necessarily be bad for Western Europe because it might counteract the global warming. According to a new study by the Leibniz Institute of Marine Sciences (also known as IFM-Geomar), no weakening of the Gulf Stream has yet been detected. Even if the Gulf Stream were to cease to flow altogether, the warming trend in Europe would merely grow somewhat weaker.[64]

And, as was mentioned above, we need not fear that the South Pole will become so warm in the next 100 years or so that the ice there will begin to melt. But if temperatures were to continue to increase far beyond that, so that one day even the Antarctic ice would melt away, sea levels could rise by 61 meters, if only after hundreds of years.[65] In that case, the Netherlands would lie nearly completely under water, and Düsseldorf, Hannover, and Berlin would become seaports. Hardly anyone dares to deal with such projections.

The Stern Commission tried to calculate the consequences of global warming in monetary terms. For the BAU scenario, according to which temperature rises to about 4.5°C above the present level (5.5°C above pre-industrial level), they estimate an annualized damage of between 5 percent and 10 percent of annualized world consumption, "now and forever."[66]

Measures to avoid and mitigate this temperature increase by reducing CO_2 emissions are also expensive, but they would still be cheaper for humanity, according to the Stern Commission. The Review estimates annualized mitigation costs of around 1 percent of annualized world consumption if the temperature increase is limited to 2°C above today's level or, equivalently, CO_2 concentration of 550 ppm.[67] Relative to the BAU scenario, mitigation of this extent would reduce the damage by between 2.8 percent and 5.6 percent of consumption (2.5/4.5 of the above-mentioned annualized damage figures for a 4.5°C increase). Thus, the annualized net gain from a mitigation strategy aimed at limiting the temperature increase to 2°C above today's level would be between 1.8 percent and 4.6 percent of annualized consumption. Even though the numerical calculations behind such numbers rest on many assumptions and are therefore not free from arbitrariness, they do provide a learned argument for why, from an economic point of view, there is a case for acting now to slow climate change. An arguably stronger qualitative case leading to the same conclusion will be presented in chapter 4.

Let us try to pack the air in sacks.

2
Reshaping the World's Energy Matrix

The first climate accords

The scientific findings regarding climate change have alarmed people around the world, unleashing a wide-ranging debate that has had the effect of gaining an increasing number of supporters for policy measures aimed at curbing CO_2 emissions through a reshaping of the world's energy matrix. This chapter gives an overview of the amount and kind of energy countries produce and discusses the options for carbon-free energy alternatives as well as the actual policy measures chosen for the purpose of moving toward more carbon-friendly energy sources. It does not yet focus on bioenergy and resource-conservation issues; those are treated in later chapters.

Though the climate debate is old, a worldwide breakthrough in public awareness of the global warming problem was made by former US vice-president Al Gore, who reached a huge audience with his 2006 film *An Inconvenient Truth* and who was (with the IPCC) a recipient of the 2007 Nobel Peace Prize. And the Stern Review, published in 2007, exerted a major impact on both science and politics. Because the Stern Commission had been appointed by Prime Minister Tony Blair, the British government accorded the report a great deal of publicity and managed to make climate change a dominant theme in Europe's capitals. At the G8 Summit in 2007 at Heiligendamm, Germany, wide-ranging proposals to battle the greenhouse effect were put forth, and those proposals were adopted in principle a year later at the G8 Summit in Toyako, Japan. Essentially, the G8 countries subscribed to the goal of halving their CO_2 emissions by 2050.[1] The hope that this goal would find support among a much wider range of countries had then been pinned on the UN Climate Conference, to be held in Copenhagen in 2009, but that conference was a failure.[2] Although the next conference, held in Mexico in 2010, achieved such a proclamation,[3] no concrete measures were agreed to and no burden-sharing plan was adopted that would help attain this goal.

Public awareness of the climate problem was preceded by more than 30 years of intensive scientific discussion, in which climate researchers raised many warnings. Scientific discussion of the greenhouse effect had already started in the nineteenth century with the works of Fourier, Tyndall, and Arrhenius. The first mention of the potential perils resulting from a man-made reinforcement of this effect is found in a study by Revelle and Suess dating from 1957,[4] but it wasn't until the first global climate conference, held by the World Meteorological Organization (WMO) in Geneva in 1978, that scientific warnings rose in a chorus. The Geneva conference is considered to have initiated the current wave of climate research, leading eventually to insights and findings such as those addressed in chapter 1 of this book. Since 1972, experts from various UN sub-agencies had devoted their efforts to the relationship between climate anomalies and the influence of human activities on climate evolution. They pointed out that CO_2 concentration in the atmosphere deserved the greatest attention of the international community because it could provoke grave changes in global climate.

Political voices first addressed the issue during the Toronto conference of 1988, calling for a 20 percent reduction in global CO_2 emissions by 2005, as well as for the formulation of an international convention on the matter. Around 300 natural science scholars, economists, sociologists, and environmentalists from 48 countries took part in the conference. In that same year, the United Nations, in conjunction with the World Meteorological Organization, established the Intergovernmental Panel on Climate Change (IPCC). The word "panel" understates the true magnitude of the globe-spanning research network that has been put together. The IPCC presented its first report at the second World Climate Conference, held in Geneva in 1990. At that conference it was agreed to start negotiations for an internationally binding agreement on climate change. In 1992, the United Nations Conference on Environment and Development (UNCED), held in Rio de Janeiro, with the participation of 178 countries, established the United Nations Framework Convention on Climate Change (UNFCC). Under this convention, the by then 189 signatories committed to reducing their carbon dioxide emissions in order to slow down climate change. The first and thus far the only commitment to concrete actions was achieved during the Kyoto conference. That was a step forward, but it was no real solution.

The Kyoto Protocol

The Kyoto Protocol, signed in 1997, opened a new chapter in climate policy by having certain countries commit, for the first time, to reducing their emissions of greenhouse gases by a certain percentage. Five years earlier, the Environmental Treaty of Rio de Janeiro had only brought commitments to cut back on CO_2 output, leaving open by how much each country would reduce its own emissions. For that reason, during the next World Climate Summit, held in Berlin in 1995, it was agreed to start negotiations aimed at devising a binding protocol that would set reduction goals and deadlines for the industrial nations. This "Berlin Mandate" was implemented in the Kyoto Protocol.

The Kyoto Protocol, ratified by 189 countries, established the goal of reducing greenhouse-gas emissions over the period 2008–2012 by 5.2 percent on average relative to the year 1990. The Kyoto Protocol was a milestone in international coordination for battling climate change. However, it was far from truly successful, since most of the signatory countries faced no consequences at all. Caps were set for only 51 countries, which together accounted for 28 percent of anthropogenic CO_2 emissions in 2005.[5] These include the 27 member countries of the European Union (which account for 15 percent of global CO_2 emissions), Russia (5.7 percent), Japan (4.5 percent), Canada (2 percent), Ukraine (1.1 percent), Norway (0.14 percent), New Zealand (0.12 percent), and Iceland (0.008 percent).

China and India have signed and ratified the Kyoto Protocol but are exempted from a cap in order not to imperil their economic development. The United States has signed the protocol, but the US Senate hasn't ratified it. Australia, whose per-capita CO_2 emissions are among the world's highest and which accounts for 1.3 percent of the global total, was among the countries that rejected the protocol at first. But Australia changed its position and accepted effective emission constraints in 2007, when Kevin Rudd of the Labor Party became prime minister. Australia signed and ratified the protocol in December of that year.

By 2010, about 27 percent of global CO_2 emissions had come under the Kyoto restrictions, a bit less than five years earlier because of the rapid increase in emissions by the unrestricted countries. For a while there was some hope that the United States, which accounts for about 19 percent of the global CO_2 emissions, would revise its position once again and ratify the protocol. That would have brought about 46 percent of the worldwide emissions under the protocol's stipulations. President

Barack Obama had suggested that the US might change its position, but unfortunately the economic crisis of 2008 and 2009 absorbed the funds and attention necessary to pursue this goal further in the short and the medium term.

The Kyoto Protocol established specific percentage reductions for the average emissions of individual countries for the years 2008–2012 relative to 1990. Japan, for instance, must reduce its emissions by 6 percent, while the EU-15 committed to a collective 8 percent reduction target, targets for the individual countries being set through a burden-sharing plan. The current EU, with 27 member countries, has no common reduction target within the framework of the Kyoto Protocol.[6] That notwithstanding, the new EU members, with the exception of Poland, Hungary, Malta, and Cyprus, have entered into individual 8 percent reduction commitments under the Kyoto Protocol. (Poland and Hungary have committed to 6 percent.)

The European Union has been pushing for much more ambitious targets in a post-Kyoto agreement, arguing that a 50 percent if not an 80 percent reduction in emissions relative to 1990 would be necessary by 2050 in order to limit the increase in the global average temperature relative to pre-industrial times to 2°C.[7] At the 2009 G8 Summit in L'Aquila, Italy, the EU members of the G8 even succeeded in making this an official G8 goal, though one without any commitments. Industrialized countries would have to curtail their emissions by 80 percent, and other countries by 50 percent.[8]

Figure 2.1 gives an overview of the legal commitments and of the actual reductions until 2008. The figure pertains to the 20 countries that account for the largest CO_2 emissions. The upper panel shows which countries are the largest polluters. The larger an economy, the higher its CO_2 emissions. Emissions don't depend solely on economy size, however, but also on the countries' individual characteristics and technological level. Until 2006, the United States, with the world's highest GDP, topped the list of CO_2 emitters, but since 2007 China, whose GDP currently is the second-highest in the world, has been first in CO_2 emissions. Japan and Germany, whose GDPs rank third and fourth, rank fifth and sixth in emissions. Russia and India, also rather high emitters, come third and fourth in the CO_2 ranking, although in terms of GDP they rank eleventh and twelfth.

As the lower panel shows, Russia, Germany, and Ukraine have achieved high percentage reductions. This undoubtedly has to do with the collapse

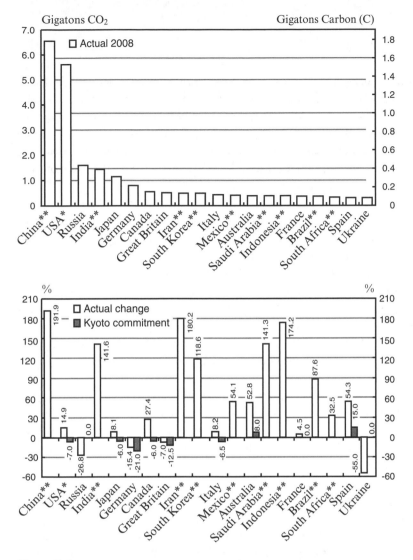

Figure 2.1
CO_2 emissions and reductions commitments of the 20 largest CO_2 emitters within the Kyoto Protocol framework, 1990–2008. *Kyoto Protocol not ratified. **Not subject to a cap. The upper panel shows emissions in 2008. The lower panel's light columns show the actual percentage change in emissions until 2008 of each country relative to the 1990 baseline year; the corresponding darker columns show the reduction committed to under the Kyoto Protocol from the 1990 baseline year to the average of the years 2008–2012 for the cases in which such a reduction was committed to. France, Ukraine, and Russia adopted no reduction commitments, but neither were they allowed to increase their emissions. While the EU as a whole committed to an 8 percent reduction per country, differing commitments were agreed among the member countries in accordance to a burden-sharing plan. The lower panel shows the commitments resulting from this burden-sharing plan. Sources: International Energy Agency, *CO₂ Emissions from Fuel Combustion*, Highlights, 2010 edition (http://www.oecd-ilibrary.org), p. 13; International Energy Agency, *CO₂ Emissions from Fuel Combustion Statistics,* Indicators for CO_2 emissions, CO_2 Sectoral Approach, 2010 edition (http://www.oecd-ilibrary.org/statistics).

of communism, which in each of those three countries led to the disappearance of a large obsolete industrial base. Worthy of note is the looming climate problem represented by the emerging economies of China, India, Brazil, Indonesia, Iran, and South Korea. The formidable economic development of these countries is made possible by a rapidly increasing consumption of fossil fuels, which brings a corresponding increase in carbon dioxide emissions.

Of all the countries represented in figure 2.1, Germany has committed to the largest reduction (21 percent). And with an effective reduction of 15.4 percent by 2008, it had come close to attaining its target.[9] (By 2009, Germany had already surpassed its reduction goal.) In addition to the demise of the former German Democratic Republic's industrial combines, the sluggish German economy at the time helped to keep emissions levels down: the slower the wheels turn, the less energy they burn. Next to Italy, Germany's economy was the most sluggish among all EU countries from 1995 to 2008, enabling it to reduce its CO_2 emissions more easily than faster-growing countries such as Canada, Spain, and South Korea.

Great Britain was also exemplary, committing to far-reaching reductions and partly achieving them by 2008. In contrast, the United States, Japan, Canada, Iran, Australia, and Spain have disappointing records, as do most of the original EU countries, some of which have increased their emissions markedly.

The climate sinners

As it is a bit unfair to compare countries by their absolute emissions, the left-hand side of figure 2.2 adds a per-capita comparison. The 13 OPEC countries, which burn lots of fossil fuels but are very small, aren't included (with the exception of Saudi Arabia). The largest per-capita emitters are, obviously, the United States, Russia, Australia, Canada, Germany, and Japan. All these countries boast high per-capita GDPs, and some have strong manufacturing sectors, so these levels aren't surprising.

It is more interesting to look at CO_2 emissions relative to GDP, as shown on the right-hand side of figure 2.2. This makes it possible to say something about how climate damaging or climate friendly the respective production processes are. Switzerland and Sweden stand out as shining examples. With only about 150 grams of CO_2 per dollar

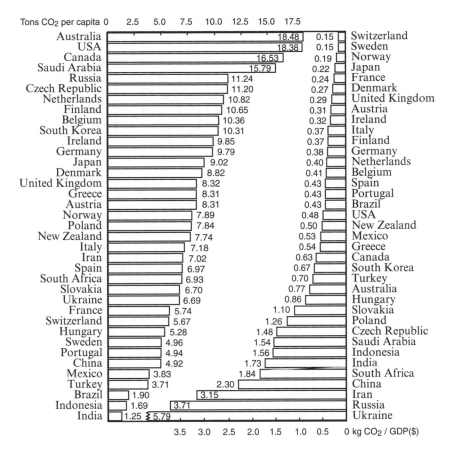

Figure 2.2
CO_2 emissions of OECD countries and of major emerging economies, per capita and in relation to GDP, in 2008. Source: International Energy Agency, *CO_2 Emissions from Fuel Combustion Statistics,* Indicators for CO_2 emissions, CO_2 Sectoral Approach, 2010 edition (http://www.oecd-ilibrary.org/statistics).

of GPD (in US dollars, at 2000 prices), these countries are world champions in the climate friendliness of their production processes. Among the bigger countries, France and the United Kingdom perform well. Hydro power and nuclear energy account for their top positions in the ranking.

The United States is in eighteenth place. The dirtiest producers are Russia and Ukraine, which have not yet managed to get rid of their communist-era industrial facilities. China, on account of its size and its blistering rate of economic development, poses a particularly difficult problem for the world. Though China recently made some efforts to develop showcase electric vehicle projects to convince the world of its willingness to curb its CO_2 emissions, the amount of CO_2 it spews out goes well beyond what would correspond to the size of its economy, and things will get worse before they improve. Currently China opens about seventy 500-megawatt coal-fired power stations per year. These power stations feed China's electricity-hungry industry and also the batteries for its electric vehicles. It is true that in 2008 China's electricity from renewable sources accounted for 17.4 percent of total electricity, a bit higher than the European Union's 16.6 percent. However, this was nearly exclusively traditional hydro power. Electric power from more sophisticated sources, such as wind and sunshine, didn't yet play any role, even though China is a leading exporter of solar panels and wind power stations, benefiting largely from Europe's feed-in tariffs for electricity from renewable energy. According to a study by the IEA, China even subsidizes fossil-fuel consumption. The prices for oil, coal, electricity, and natural gas it charges their final consumers are below world market prices. Industrialized countries do the opposite: they levy taxes that increase the prices of fossil fuels against those prevailing in the world market.[10] To be fair, however, one should acknowledge that Europe and the United States have thus far enjoyed carbon-intensive Chinese products that they imported cheaply.

Thanks to their relatively low level of industrialization, China and India still exhibit comparatively low per-capita carbon dioxide emissions in absolute terms, but that could change very quickly. Since the end of communism, both countries' economies have grown at high rates, often exceeding 10 percent per year, a development that translates into increased levels of CO_2 emissions. Though China accounted for only 6.4 percent of the world's GDP in 2008, it emitted as much CO_2 as Russia, India, Japan, Canada, Great Britain, Iran, and Germany combined, accounting for about 22 percent of anthropogenic CO_2 emissions.

The world's energy matrix

What scope is there for non-fossil fuels to replace fossil fuels? Many believe the future economy will be carbon-free, with all energy coming from sunlight, wind, and water; others place their hope in nuclear power.

To get a feeling for the enormous effort that might be required to make these dreams come true, it is useful to first get an overview of the actual energy flows in the developed countries.

Apart from the direct absorption of sunlight by the environment, 35 percent of the primary energy sources (fossil fuels, renewable energy, and nuclear power) processed by mankind is used for the direct production of heat (by burning), and 59 percent to power all kinds of engines, including power plants and combustion engines. Six percent of primary energy is used not for energy production but as an input in the production of chemicals. Of the 59 percent used for engines, about 38 percentage points go into the production of electricity, 20 percentage points are used for airborne, water, and overland transportation; the remaining percentage point is used to power other machines that operate with primary energy—for example, gas turbines in pumping stations or stationary diesel engines.[11]

The primary energy is converted into final energy in the forms of refined fuels, electricity, heat, kinetic energy, and the like. Figure 2.3 shows where the final energy consumed comes from. It applies to the OECD countries and refers to 2007, the most recent year for which data are available at this writing.

The figure shows that crude oil is by far the largest source of final energy consumed in the world, contributing 46.1 percent of the total, followed by natural gas, with 26.7 percent, and coal, with 13.1 percent. While coal is primarily used for the production of electricity, crude oil is primarily used for fuel production, and natural gas primarily for heating. Together, these three fossil energy sources account for 85.9 percent of the OECD countries' energy consumption.

Nuclear energy accounts for 5 percent of the total final-energy use. Renewable energy, on which the world's hopes are pinned, accounts for 8.7 percent. By far the largest source of renewable energy is bioenergy, of which wood is still the main item. Bioenergy contributes nearly as much as nuclear energy does (4.76 percent).

A word of caution is appropriate, however, when interpreting the above figures, as final energy is not the same as primary energy. "Final energy" refers to the energy that is available to the end user. "Primary energy" refers to the energy used as input to produce the final energy. Since the efficiency factors of the various technical processes differ, statistics on primary energy and on final energy produce different shares for each energy carrier. If the final energy is consumed in a form similar

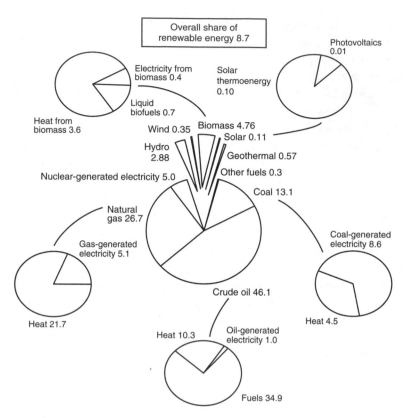

Figure 2.3

The final-energy matrix of the OECD countries (percentages in 2007). This figure shows the principal end uses of each energy carrier, grouped in relation to the primary energy carrier from which they were produced. For example, crude oil is used to produce automotive fuels, electricity, and heating fuel. The "other fuels" include the input of waste-incineration plants. Sources: International Energy Agency, *Energy Balances of OECD Countries*, 2009 edition, p. II.13; *OECD Total: 2007*, Contribution from Renewable Energies and Energy from Waste, p. II.206; Ifo Institute calculations.

to that of the primary energy, the efficiency factor is usually high; such is the case for petroleum products or natural gas, for example. If, conversely, the final energy is consumed in a strongly altered form, such as electricity, the efficiency factor is small. In a nuclear reactor, for instance, only about one-third of the heat energy generated is transformed into electricity. In a coal-fueled power station, the efficiency factor is between 35 percent and 46 percent, depending on the technological vintage of the station. A modern diesel engine transforms about 40 percent of the primary energy contained in the fuel into kinetic energy. Some modern condensing boilers for gas or oil, such as those found in many cellars, have efficiency factors of nearly 100 percent. That is why it is much cheaper to heat a home with a furnace than with a fan heater running on electricity generated from fossil fuels.

Figure 2.3 refers to final energy. A statistic for the corresponding primary energy would see the contribution of coal increase from 13.1 percent to 21.1 percent, and the contributions of crude oil and natural gas shrink from 46.1 percent to 39.9 percent and from 26.7 percent to 22.9 percent, respectively. The contribution of nuclear power, on the other hand, increases from 5.0 percent to 10.8 percent.[12]

Note also that the reference to final rather than primary energy use implies that 34.9 percent of the final energy comes from liquid fuels. This seems to contradict the initial statement that 20 percent of world's primary energy is used for transportation, but it doesn't. For one thing, figure 2.3 refers to OECD countries rather than the world. For another, it is again the distinction between final and primary energy use that explains the difference. The OECD countries' share of liquid fuels in the primary energy used, which is roughly equivalent to transportation, is about 24 percent. And the world's share of liquid fuels in final energy consumption amounts to 30.6 percent.[13]

The picture is again different in the case of electricity generated from waste, sludge, or biomass, whose contribution in the final energy statistics is usually measured according to the substitution method, which takes into consideration the average thermal efficiency that would have been generated in a coal-fired power plant to produce the same amount of electricity. These peculiarities explain why "green" electricity has a share of 2.7 percent in the primary-energy statistic but is represented in figure 2.3 as having a share of 3.8 percent. Overall, renewable energy accounts for 8.7 percent of the final energy, while its share in the primary-energy statistic is 6.5 percent.

"Green" electricity

In view of the great hopes that many people place in a carbon-free energy future, with electricity generated by wind and sunlight, and of the enormous efforts that many countries have made in this regard, it is a bit disappointing to see that both of these sources of renewable energy contribute so little to final energy consumption. While figure 2.3 shows that 86 percent of the world's final energy consumption is currently obtained from fossil fuels, electric power generated from wind and sunlight contributes not even 0.4 percent, and photovoltaic energy only 0.01 percent. There is obviously a long way to go before these energy sources are able to make more than negligible contributions.

For the time being, hydro power is much more important than solar or even wind energy. Hydro power has long been the world's most important kind of electricity from renewable energy sources. After the first hydro power station was built in Northern England in 1880, similar stations started sprouting around the world. Today, hydro power accounts for nearly 3 percent (2.88 percent) of the world's total final energy use and for 16 percent of the world's electricity production.

The reason for the poor performance of solar and wind energy thus far is that these forms of energy, though in rich supply by nature, are thinly spread out and hence difficult to collect and concentrate in a particular location. The grid necessary to collect and concentrate these sources of energy from huge areas of land or water involves extremely high financial investments. The energy contained in the rain falling from the sky is similarly thinly and widely dispersed. However, through a system of creeks, rivers, and lakes, nature concentrates this energy in particular locations, making it comparatively easy to collect the energy there.

Austria, Sweden, and Switzerland enjoy particularly large hydro resources. On average, electricity from renewable sources accounts for 15 percent of the final energy consumption and 57 percent of the electricity consumption of these countries, of which 53 percentage points are accounted for by hydro power. In Norway, which has many fast-flowing rivers rushing down from its mountains, renewable sources account for 57 percent of the total final energy consumption and for 99.6 percent of electricity consumption, of which 98 percentage points are hydro power.[14]

In Germany, on the other hand, renewable energy thus far accounts for only 3.3 percent of total final energy consumption and 14 percent of electricity consumption.[15] This is surprising insofar as in 2009 Germany had already installed 21,000 wind turbines and more than half a million photovoltaic roofs and other photovoltaic appliances.[16] The obvious explanation for these differences is that the former group of countries has ideal geological conditions for hydro power, while wind and solar energy are ineffective by comparison. Although the German photovoltaic appliances contributed only 0.2 percent to the final energy provided and 1 percent of electricity, the German public is deeply convinced that this is the perfect replacement technology for fossil fuels.[17]

In the United States, despite all the efforts, the role of renewable energy is currently also limited. Inland from San Francisco thousands of wind turbines whirr away, providing electricity for the city. In the morning, the sea is warmer than the land, and the wind blows toward the ocean. In the afternoon, the picture reverses itself. Thus, precisely at the time when people are at home, cooking meals, the wind blows and provides an ideal supply of energy. And, being in the middle of a desert, the wind turbines don't annoy anyone. Nevertheless, in the United States renewable energy accounted for just 6.7 percent of total final energy consumption and 8.3 percent of electricity consumption in 2009. Wind electricity itself accounted for 0.2 percentage points of final energy consumption and 0.8 of electricity use.[18]

Although renewable energy has only a limited role in the United States, the country's size makes it the world's largest producer (and consumer) of wind power, accounting for 24.1 percent of the world's production. It is followed by Germany, with 18.6 percent, Spain with 14.5 percent, and China with 8.0 percent. This is shown in figure 2.4, which also gives the respective country shares for the other non-fossil sources of electric power: solar power, hydro power, electric power from biomass, and nuclear power.

Germany, on the other hand, is the world's largest producer of photovoltaic electricity, followed by Spain and Japan. In view of the small overall volume of solar power, this championship isn't particularly important, but it does reveal the idiosyncratic political preferences Germany has developed, which have led to voluminous public subsidy programs for renewable energy, the most important of which are feed-in tariffs. Feed-in tariffs are publicly guaranteed prices for the "green"

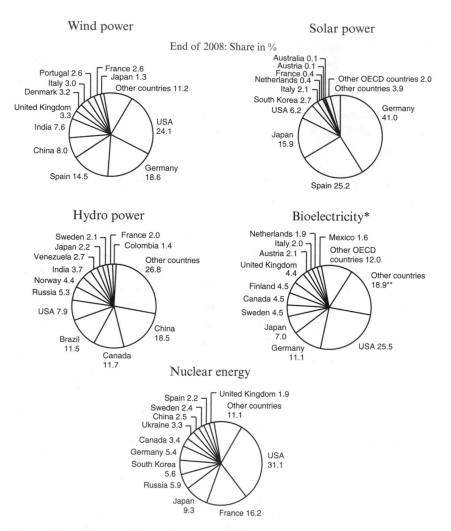

Wind power

Solar power

End of 2008: Share in %

Hydro power

Bioelectricity*

Nuclear energy

Figure 2.4
World market shares in the production of non-fossil-fuel-generated electricity (2008). *Electricity from wood, biogas, and biofuel. **For bioelectricity, 2008 data were estimated on the basis of 2007 data. Sources: International Energy Agency, *World Energy Outlook 2009*, 2009; IEA, *Energy Balances of OECD Countries*, 2009; IEA, *Energy Balances of Non-OECD Countries*, 2009; IEA, *Electricity Information 2009*, 2009; *EurObserv'ER* no. 6, March 2010, Wind Energy Barometer; Renewables Global Status Report 2009 update, REN21 Renewable Energy Policy Network for the 21st Century. Hydro power: *BP Statistical Review of World Energy*, June 2009; Nuclear: International Atomic Energy Agency, *Nuclear Power Reactors in the World*, 2009 edition, Vienna 2009; Biomass: IEA, *Electricity Information 2009*; Ifo Institute calculations.

current produced that are way above the market price, necessarily combined with a purchase obligation on the part of grid owners.

The German policy measures can be attributed largely to the rise of the Green Party, a small party, currently boasting only 49,000 members, that was founded in 1980, emerging out of protests against nuclear weapons. Nevertheless, with 11 percent of the popular vote in Germany's last parliamentary election, it is probably the world's strongest environmentalist party. Directly and indirectly, via other parties poaching their initiatives in order to keep their own voters, the Greens have succeeded in reshaping Germany's energy supply toward renewable sources, making Germany the world's laboratory for a future free of fossil carbon. A plan to set up a gigantic solar power complex in the Sahara to generate electricity for African and European countries is only one of many ambitious projects on which Germans pin their hopes.

As figure 2.4 shows, Spain is currently number two among the world's producers of solar power, with a market share of about 25 percent. This is due only in part to its ideal weather conditions. More important is the fact that Spain has also adopted a feed-in tariff system that makes solar power privately profitable. As was mentioned earlier in the book, Spain not only has photovoltaic panels on many roofs; it also runs big solar power stations at which concentrated sunlight is used to activate steam turbines.

Japan, like Germany, has negligible domestic energy sources and therefore has supported the production of solar energy. In Japan, a household pays only for the net power consumed. Utilities have installed electricity meters that run backward if a household feeds self-produced electricity into the grid. This certainly accounts for the large number of solar panels in Japan, which make the country number three in solar power production. Japan's climatic and geographic conditions ensure that only end users operating rooftop solar panels can profit from the backward-running meters. Furthermore, Japan promoted rooftop solar panels through a special government program.

Surprisingly, wind power is considerably less important than solar power in Japan. Perhaps the frequent typhoons make wind power somewhat risky, as the turbine blades would have difficulty coping with powerful gusts, or maybe it is the steeply falling seabed that makes offshore installation impossible. Still, despite its large share of the world market in solar power and its small share of the world market in wind power, Japan still produces more electricity from wind than from sunlight. This also goes to show how limited the contribution of solar power really is.

Figure 2.4 illustrates how essential geographical conditions are for hydro power. However, Norway, Sweden, Switzerland, and Austria, where this form of energy is relatively important, don't play much of a role in a worldwide context, on account of their small size. Large countries such as China, Canada, and Brazil dominate this form of energy supply.

Bioelectricity plays a more important role in non-fossil-fuel production of electricity than is commonly realized. As figure 2.3 shows, bioelectricity is, from a global perspective, close in quantitative importance to wind electricity, and much more important than photovoltaic electricity. Bioelectricity is electric power produced in small power stations, often attached to farms, by burning wood or converting biomass into methane by way of fermentation, and burning the methane to run electric power generators. As figure 2.4 shows, the United States produces about one-fourth of the world's bioelectricity, followed by Germany, which produces about 11 percent.

The nuclear alternative

Although nuclear energy isn't renewable energy in a strict sense, it doesn't use fossil fuels, and it produces electricity without emitting CO_2 (except, of course, for emissions in the investment phase and negligible emissions from auxiliary services). This is why it is also included in figure 2.4, which shows that the United States produces 31.1 percent of the world's nuclear power and France 16.2 percent. Japan, Russia, South Korea, and Germany have minor but still significant market shares.

The current revival of nuclear power was triggered by the dramatic increases in the prices of fossil fuels in the years before the global financial crisis, culminating one day in July 2008 in a price of $147 per barrel of crude oil, more than seven times the average price in the 1990s.

Figure 2.5 gives an overview of existing and planned nuclear plants. In 2010 there were 440 nuclear power stations in the world, 59 new stations were under construction, 149 were firmly planned, and there were preliminary plans for another 344. The European Union currently plans to build or is building 49 new reactors. China leads the field, with

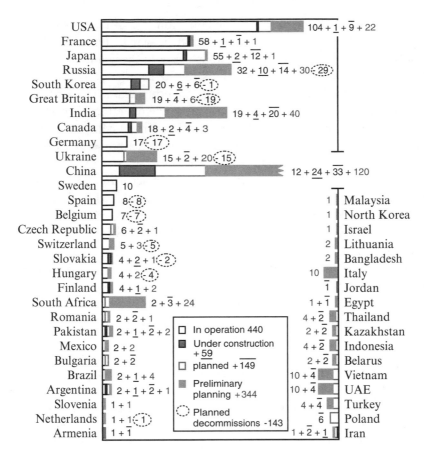

Figure 2.5
Nuclear plants around the world in 2010. Sources: World Nuclear Association, *World Nuclear Power Reactors and Uranium Requirements* (http://www.world-nuclear.org), January 2011, and national reports.

24 new stations under construction and another 153 planned. China currently opens three one-gigawatt nuclear power stations per year.

Even before the accident at Fukushima in March of 2011, Germany, Spain, and Belgium had planned to give up nuclear power entirely. Germany plans to shut down all 17 of its nuclear power plants. (Five plants in the formerly communist east were closed in the early 1990s.) Though the date for Germany's "nuclear exit" was hotly debated, the decision to do so was supported by all five political parties that hold seats in the German parliament. After Fukushima, Germany accelerated its plans. The German government immediately switched off seven of its nuclear power plants, which produced 7 percent of the country's electricity, and agreed to shut down the others by 2022. In contrast, with the exception of Spain and Belgium, nearly all other countries producing nuclear energy declared that they would continue operating their plants, although most promised to invest in more safety. This reiterates the point, made above, that under the influence of the Green Party Germany seems to be choosing an idiosyncratic path that sets it apart from nearly all other countries. Insofar as German and Austrian scientists (Otto Frisch, Otto Hahn, Lise Meitner, and Fritz Strassmann) developed the theory of nuclear fission, which is the basis of all nuclear power stations and for which Hahn was later awarded the Nobel Prize, this is a remarkable development.[19]

Nuclear energy undoubtedly involves particular risks. At Fukushima in March of 2011, an unusually strong tsunami destroyed the cooling systems of five reactor blocks. Four of the six reactor blocks exploded, and the cores in three of them partially melted. At this writing, a month after the Fukushima disaster, it is estimated that about 18,000 people have died from the Japanese tsunami, but no fatalities have as yet been attributed to the nuclear catastrophe itself. (Severe long-term consequences for the rescuing personnel can be expected.) At Three Mile Island in 1979, a mishap led to partial melting of a reactor's core, but there were no fatalities and no injuries. At Chernobyl in 1986, a reactor overheated, setting off a hydrogen explosion that killed many people. How many is vigorously disputed. The International Atomic Energy Agency, the World Health Organization, and the United Nations Development Program reviewed the available studies, submitted them to a scientific quality test, and published a joint statement on the matter. According to their findings, about 50 people died directly as a consequence of the Chernobyl accident, and the long-term effects could eventually bring the death toll to 4,000.[20] The WHO also examined contaminated areas farther away from the accident site. In those areas, in the long term, a further 5,000 people may die.[21]

The catastrophes are grim, no doubt. They call for urgent safety checks based on agreed international standards. Many reactors may need additional safety devices; others may even have to be shut down. It is important, though, for policy makers to understand some basic differences between the different reactor types before they take hasty actions as in Germany.

A nuclear reactor needs a moderator to keep the chain reaction going. Whereas the Chernobyl reactor used graphite as a moderator, nearly all modern reactors used in the West, and nowadays also in Russia, use normal (light) water. The Fukushima reactors also belonged to that category, though they were of an older type with only one rather than two water circuits from the core to the turbines (a boiling-water reactor). Unlike graphite, water expands and leaks when a reactor begins to overheat, automatically interrupting the chain reaction, at least for a while. Whereas at Fukushima the chain reaction was interrupted when the tsunami hit, the Chernobyl catastrophe resulted from an uncontrolled chain reaction caused by a misguided experiment in shutting down the reactor. Despite the interruption of the chain reaction, three of the Fukushima reactors exploded because residual heat had accumulated after the cooling system was destroyed.

Many people believe that there were nuclear explosions, similar to those caused by nuclear bombs, at Chernobyl and Fukushima. That is not true. For a nuclear explosion to occur, two highly enriched non-critical metallic masses of uranium 235 (or plutonium 239) must be brought together at a very high speed in a highly compressed state in order to assemble them into a critical mass.[22] This is not possible in any nuclear power plant, anywhere. What did happen was that, because of the overheating of the core, the metal casing containing the rods for controlling the nuclear reaction reacted with water and produced hydrogen, which then exploded and destroyed the reactor buildings. Because the Chernobyl reactor had no containment vessel covering the core, the explosion immediately set free tons of radioactive material, and then burning radioactive graphite continued to evaporate into the atmosphere for days. At Fukushima, the radioactive material largely stayed within the containments. However, security valves set some of the contaminated water vapor from inside the containments free, and contaminated water leaked for a while from one of the five containments. At this writing, the Fukushima reactors are still being cooled provisionally from outside, and it is still not clear whether and when it will be possible to reinstall permanent cooling systems. Permanent cooling is necessary to prevent a meltdown of the core, because after such a meltdown the fissile material may become so condensed that a new chain reaction will start. The Chernobyl and Fukushima accidents have been perceived as equally severe by many observers, but six weeks after the accident ten times as much radiation had been released to the environment at Chernobyl as at Fukushima. The Russian atomic energy agency, ROSATOM, therefore argued that Fukushima did not deserve classification as a catastrophic accident comparable to Chernobyl, as the Japanese atomic energy authority had argued it did.[23]

The events show that, apart from redundant and earthquake-proof emergency cooling systems, it is essential for the safety of nuclear reactors to be able to keep the melting of the core under control. Europe's biggest reactor, currently under construction in Olkiluoto, Finland, even has a "core catcher," a flat ceramic pan that would prevent a chain reaction even after a meltdown of the core. It also has an internal supply of cooling water sufficient to cool the melted material down to a level where a chain reaction is excluded, and a double-wall containment vessel that can withstand airplane attacks. A similar reactor is under construction in Flamanville, France.

Radioactivity is an issue of concern for many. However, it is a natural phenomenon that has accompanied life from its very beginning, as the Earth consists of radioactive material that has been decaying and radiating since the planet originated. Natural radioactive decay is responsible for about 62 percent of the Earth's warmth.[24] Uranium and a series of other elements that occur in the Earth's crust (mainly thorium 232 and potassium 40) decay and emit radioactivity, just as a fuel rod in a nuclear reactor does. People who heat their homes with geothermal energy are in effect making use of nuclear energy. In locations where uranium has concentrated naturally and water has come into play, acting as a moderator, chain reactions have occurred in the past. At Oklo in Gabon, such a natural reactor got going 2 million years ago and was active for half a million years.

Usually, though, uranium decays without triggering a chain reaction. Because the half-life of U^{235} is 700 million years, this process goes on for a very long time, albeit with ever-lower intensity. When the Earth came into being, many more radioactive substances existed, possessing much shorter half-lives and therefore decaying more rapidly. Plutonium was one such substance. As a result of its relatively short half-life (approximately 24,000 years), there is practically no naturally occurring plutonium anymore. Also extremely rare is U^{234}, which has a half-life of 245,500 years—a blink of an eye in geologic terms.

Life arose on Earth about 3.5–4.5 billion years ago. Bathed in radiation, the cells making up all living organisms had to learn to develop effective mechanisms by which to repair radiation-induced genetic damage. These mechanisms manage today's low residual radiation easily. They were built to cope with much heavier bombardment, since at the time when life arose on our planet cells were subjected to five times as much radiation as today.[25] Radiation may cause harm to cell nuclei by damaging their genetic information. But organisms, humans included, have learned to cope with this. It sounds unbelievable, but about 50,000 gene defects are caused *in each cell* of a human body *daily* by radiation and other causes.[26] Nevertheless, our bodies easily handle these defects and repair them immediately.[27]

The extra radiation coming from nuclear reactors is negligible in this regard. At a distance of 50 kilometers, a coal power plant generates three times the radiation of a nuclear power plant. At the fence surrounding a nuclear power plant, the radiation is only 0.25 percent higher than the average radiation on the Earth's surface—much less than the natural variation between different locations. Thus, if a Swiss leaves his rocky, radioactive environment to live directly at the fence of one of the nuclear reactors in northern France, he reduces his exposure to radioactivity by 50 percent.[28] This doesn't imply, of course, that radioactive waste involves no danger and doesn't have to be disposed of.

Storing fossil-fuel waste and nuclear waste

The difficulty of finding safe ways to store nuclear waste is one of the major arguments that environmental activists raise against nuclear power. To understand the problem, we must recognize the difference between weakly and strongly radioactive substances—that is, between low-level and high-level waste. High-level waste consists mainly of spent nuclear fuel, vitrified fission products from reprocessing, and activated core components. Low-level waste consists mostly of operational waste from nuclear plants (ion exchangers from cooling fluid purification systems, filters, vaporizer concentrates, and the like) and radioactive debris from dismantled nuclear plants. The most problematic high-level waste is plutonium. Its 24,390-year half-life makes it dangerous for human beings practically forever.

The problem of storing plutonium can be reduced if the spent fuel is reprocessed to capture the radiating plutonium, Pu^{239}, and use it to produce MOX fuel rods, which can be used as substitutes for normal uranium fuel rods. Uranium doesn't occur in nature as a pure metal, but there are more than 200 varieties of uranium ore, the most important being uraninite and coffinite. Natural uranium comprises two isotopes (that is, variants with a different number of neutrons in the atomic nuclei): U^{235} and U^{238}. Only the former, which accounts for about 0.7 percent of natural uranium by weight, is fissionable, but its proportion must be raised to at least 3 percent in order to make it suitable for pressurized-water reactors. Otherwise no chain reaction can occur. In order to achieve such a concentration, U^{235} must be enriched. This is accomplished by removing part of the U^{238}. The enriched uranium is brought into gaseous form, then turned back into a solid from which fuel rods are made. As a rule, the fraction of U^{235} is enriched to 4 or 5 percent to produce fuel rods. When these rods are used in a reactor, some of the U^{238} captures neutrons and is converted to plutonium. The plutonium added to the MOX fuel rods by way of reprocessing the nuclear waste is also fissile and can, up to a point, replace the U^{235}. Reprocessing therefore has the double advantage of reducing nuclear waste and increasing the energy that can be generated with the uranium extracted. The United States, France, Great Britain, India, Russia, and Japan are known to have the necessary facilities, but owing to political resistance against the transport of spent fuel they aren't always used to their full capacity. Nevertheless, reprocessing can increase the energy generated from a given amount of uranium ore by a factor of 1.4 or even 1.5, and can reduce the high-level waste to around 7 percent. Even more efficient is fast breeding, which not only would extend the running times (that is, the time elapsing until exhaustion of the available stocks underground if the flow of extraction were to continue at today's pace) of uranium resources by a factor of 60 but also would reduce the high-level waste in equal measure.

Low-level waste, however, must also be stored. This encompasses about 90 percent of the radioactive waste from nuclear plants, but it is far less dangerous than spent fuel and thereby easier to store. As its radioactivity is only a fraction of that of high-level waste, the safety standards associated are somewhat lower. Some of this waste is reused as a shield for waste containers and the like.

The best place to store plutonium and other high-level waste, if reprocessing is not available, is deep underground. Granite caverns are safe and have the advantage of remaining accessible should future generations decide to reprocess the nuclear waste. Salt domes, on the other hand, can conduct heat much better than rock formations, and their plasticity ensures that no fissures or gaps will open up through which radioactive material can leak out. However, they make later reprocessing of the nuclear material difficult. In any case, the stored nuclear waste is a burden that today's generations place on many future generations. Whereas the low-level waste will no longer be dangerous after a few decades, the high-level waste will remain dangerous for thousands of years. In this it resembles the long-lived harmful effects of the waste from burning fossil fuels, namely carbon dioxide. Because a significant portion of it will not be absorbed by the oceans or the biomass, but instead will remain in the atmosphere for a very long time, contributing to the green-house effect, mankind will have to put up with the resulting higher temperatures and the damage they cause for the foreseeable future. The mean permanence of this carbon dioxide in the atmosphere is about 30,000 years, a time span comparable to that of plutonium.[29] The final repository for the waste resulting from burning fossil fuels is the air we breathe. This final storage problem has certainly a greater relevance for human safety than the nuclear waste problem.

Nowadays there is the possibility of "scrubbing" carbon dioxide from the exhaust produced by burning fossil fuels, liquefying it, stuffing it into suitable containers, and storing it underground. This is known as seques-tration. The word was used in olden times to mean the bringing into custody of a malefactor. It might be assumed that there is enough room available in the Earth's crust for such sequestration. After all, we could just pump the oxidized carbon back to where we extracted it from. One could picture, in an extreme case, a pipeline system routing CO_2 from the end users back to the originating oil reservoirs and coal mines. It would amount to extracting fuels for a short time, gaining energy from them, and pumping the rest back in. Unfortunately, it is easier said than done. The problem stems from the fact that when fossil fuels are burned, each carbon atom acquires two oxygen atoms as companions, and these must also be disposed of. Table 2.1 provides an overview of how much room is needed to store the CO_2 emissions resulting from burning various fuels. The volume of liquid carbon dioxide is expressed in relation to one cubic meter of the original fossil fuel. The results are sobering. According to the table, the resulting liquid carbon dioxide usually occupies a much larger volume than the original fuel volume. Burning a cubic meter of lignite (brown coal), for instance, gives rise to 1.4 cubic meters of liquid

Table 2.1
Volume of liquefied CO_2 at 20°C (68°F) and 55 bars (798 psi) generated by 1 cubic meter (35.3 cubic feet) of fossil fuels. Densities for anthracite, gaseous methane, and lignite from H. Recknagel, E. Sprenger, and W. Hönmann, *Taschenbuch für Heizung und Klimatechnik 1992/93* (Springer, 1992), p. 76; density for crude oil from *Dubbel—Taschenbuch für den Maschinenbau,* seventeenth edition, ed. K.-H. Grote and J. Feldhusen (Springer, 1990). Emissions factors for CO_2 emitted per energy unit from http://www.dehst.de.

Anthracite	5.4 m^3
Lignite (brown coal)	1.4 m^3
Crude oil	3.7 m^3
Methane (liquid)	1.6 m^3
Methane (gaseous)	0.002 m^3
Methane (frozen, methane hydrate)	0.4 m^3

Note: A cubic meter of anthracite weighs 1.4 metric tons. If burned, it will release about 4 metric tons of CO_2, which, owing to its density in liquid form (0.74 metric tons per cubic meter), has a volume of about 5.4 cubic meters. One cubic meter of lignite releases about 1 metric ton of CO_2, a cubic meter of crude oil 2.7 metric tons, a cubic meter of liquid methane 1.18 metric tons, a cubic meter of gaseous methane about 0.0017 metric tons and a cubic meter of frozen methane, including the frozen water with which it is mixed, 0.4 metric tons. A conversion using the specific weight of liquid CO_2 produces the volumes of liquefied CO_2 shown in the table.

carbon dioxide; a cubic meter of high-grade anthracite coal to 5.4 cubic meters of liquid CO_2, a cubic meter of crude oil to about 3.7 cubic meters of liquid CO_2, and a cubic meter of liquid methane to 1.6 cubic meters of liquid CO_2. A cubic meter of methane ice (methane hydrate) translates into 0.4 cubic meter of liquid CO_2, but there is hardly a safe way to use as storage the sites from which it was extracted, which are typically on the continental shelf, deep under the ocean surface. Only gaseous methane leaves enough room behind to store the CO_2 it releases when burned, thus providing accessible storage sites.

Thus, in most cases pumping liquefied carbon dioxide back to the original reservoirs will not be enough. This is particularly so in the case of coal. Current CO_2 scrubbing processes operate only in association with coal-fired power plants, as they require so much energy that it can only be obtained directly from such power plants. CO_2 is first scrubbed by chemical solvents that have to be heated and cooled by turns. Once the CO_2 has been isolated, it is compressed by pumps into liquid form.

In the case of anthracite coal, which accounts for 54 percent of the electricity supply worldwide, the table shows a very inauspicious volume ratio. The decommissioned coal mines are sufficient for barely a fifth of the liquid CO_2 volume resulting from burning the coal they contained originally. When it comes to lignite, the ratio appears more favorable, as you need a larger volume of lignite per unit of energy than of anthracite. Theoretically, then, burning lignite should leave enough room behind for storing its resulting CO_2. Unfortunately, lignite is extracted from open-pit mines, so no underground storage sites are left behind. (The same applies to most American anthracite coal, since it too is extracted from open-pit mines.)

If the current mines don't provide enough room for CO_2 storage, perhaps mines and gas deposits depleted in earlier times could be used. Gas deposits could be expected to offer particularly large storage volumes, as conventional methane is not a liquid or a solid but a gas. As the table shows, these deposits offer 500 times as much volume (1/0.002) as is required by the CO_2 that such methane generates. But, alas, the available volume doesn't suffice to store the CO_2 resulting from burning other fossil fuels, even in liquefied form. According to a study conducted by the IPCC, worldwide there is room for storing a maximum of 900 gigatons of CO_2 in old gas and oil reservoirs, and a maximum of 200 gigatons of CO_2 in coal mines.[30] That amounts to 1,100 gigatons of CO_2, equivalent to approximately 300 gigatons of carbon. But the total fossil-fuel resources amount to the equivalent of 6,500 gigatons of carbon. Thus, the available storage room is sufficient for barely a twentieth of the resulting liquefied CO_2. In a business-as-usual scenario, about 900 gigatons of carbon are forecast to be burned until the middle of this century.[31] The available storage places would be filled long before then.

This fact has shifted hopes toward porous rock layers and natural salt domes that are now filled with water, which would first have to be pumped away. According to the IPCC study, at least 1,000 gigatons of liquid CO_2, equivalent to about 270 gigatons of carbon, could be stored in such places. Together with the depleted oil and gas reservoirs, this would provide room for a total of nearly 600 gigatons of carbon, which would still not see us to the mid-century mark.

It is an open question whether room can be found for more than 1,000 gigatons of carbon. The IPCC's highest hypothetical estimate sees room for 10,000 gigatons of CO_2, but it is accompanied by a high degree of skepticism. Even if this volume were to be found, it would suffice for only 2,700 gigatons of carbon. Together with the upper estimates for the depleted gas and oil fields, plus the coal mines, this would amount to at most 3,000 gigatons of carbon that could be stored, equivalent to barely half of the overall carbon resources. It is not enough, but it is not little. Regrettably, there are a few other problems that significantly dent the attractiveness of this technical fix.

Consider the coal mines around Sheffield, England, or those in Germany's Ruhr district. After 200 years of mining, the underground there looks like a gigantic piece of Swiss cheese, with room enough to last a good while as a storage site if the water that currently fills the shafts and galleries is pumped out. But it actually offers no solution, because the highly interconnected mine shafts and galleries lie partly below porous ground layers and are therefore practically impossible to make airtight. And airtight all storage sites must be, not only because of climate concerns but also because CO_2 is not innocuous. Chemically non-poisonous, it doesn't harm humans directly. We even produce it in our own bodies when we burn nutrients. But it is dangerous if present in concentrations high enough to displace oxygen. Heavier than oxygen, it tends to concentrate at ground level when the air is still. If a leak should develop in a storage site, high-pressure gas could escape and displace oxygen, causing death by asphyxiation. Detailed calculations have been made for salt mines, where the situation is similar. In a salt mine of typical capacity (7–12 megatons of CO_2), under windless conditions a leak could give rise to a temporary 10-meter-thick (33 feet) carbon dioxide layer that would wipe out all human and animal life in the area affected.[32] This danger is particularly acute in valleys.

Experimental evidence of this has already been provided by nature. In 1986, carbon dioxide spontaneously released in a volcanic area near Kenya's Lake Nyos caused the asphyxiation of 1,700 people. In view of this, the idea of using old coal or salt mine galleries for carbon dioxide storage is far-fetched.

Thus, liquefied carbon dioxide can't be pumped into the ground just anyplace where room is available. Potential storage sites must be chosen in areas far from human habitation, with as much care as that required in choosing nuclear waste storage sites. In many countries, sites could surely be found to last a few years, providing a placebo to stanch the fears of the population but not sufficient to offer a real solution.

What is needed is a new pipeline system that would deliver liquefied carbon dioxide back to the former reservoirs of fossil fuels in the unpopulated wastes of Siberia, Kazakhstan, and the Arab countries. This system will have to be much larger than the existing pipeline net bringing fossil fuels to the consumer countries, as the oxygen attached to the carbon atoms will also have to be transported to the storage sites.

The fact that the oxygen would be pumped into the ground with the carbon, and thus removed from the biological cycle, poses no problem for life on Earth. Relative to the amount of oxygen in the atmosphere, the amounts of oxygen to be disposed of are so tiny that their loss is irrelevant to humans and animals. At present, oxygen accounts for 20.946 percent of the atmosphere. If all the carbon resources (about 6,500 gigatons) were to be burned and the exhaust gases released into the atmosphere, the share of oxygen in it would decrease to 20.023 percent, taking into account that photosynthesis would detach part of the oxygen bound to carbon atoms in CO_2 and then release it back into the atmosphere.[33] Should the entire stock of CO_2 produced by exhaust gases be sequestered, denying photosynthesis the chance to liberate some of the oxygen atoms, the share of oxygen in the atmosphere would fall to 19.629 percent. The decrease is a bit larger than without sequestration, but these differences are so tiny that they fade into irrelevance for living organisms. Not irrelevant, however, are the greenhouse effects caused when carbon dioxide is blown into the atmosphere, as such effects will occur despite the fact that CO_2 accounts for a vanishingly small 0.038 percent of the atmosphere. In addition, the logistical problems posed by sequestration aren't exactly irrelevant. Even if only a minuscule part of the atmosphere is involved, the task is simply gargantuan for mere humans.

In any case, the problem of finding a final repository for fossil-fuel waste is much like the problem of finding one for nuclear waste. If we want to take advantage of sequestration and stop using the atmosphere as the final repository for CO_2, sites for storing liquefied carbon dioxide will have to be managed for much longer than sites for storing nuclear waste, as liquefied carbon dioxide shouldn't be allowed to volatilize if one wants to prevent it from contributing to the greenhouse effect. Whereas nuclear waste loses its radioactivity as warmth, carbon dioxide stored in underground caverns remains exactly as it is. Mankind will have to keep the storage sites intact forever, at least for as long as it needs to protect itself from the greenhouse effect.

But the *amounts* of CO_2 that will have to be stored are a thousand times as problematic. They are so gigantic that managing nuclear waste appears trifling by comparison. Why this is so can be illustrated with a simple calculation. Let us assume that we replace a 1,225-megawatt nuclear plant with a coal-fired plant of the same capacity. We will then have to remove 11.2 million cubic meters (14.6 million cubic yards) of liquefied carbon dioxide per year.[34] This calculation doesn't even take into account that scrubbing and compressing carbon dioxide into a liquid would consume one-third of the energy generated by that coal. In reality, then, the amount of liquefied carbon dioxide is 14.9 million cubic meters. If we wanted to transport this volume by train, we would need around 6,000 forty-car trains with each tank car holding 2,200 cubic feet of liquefied carbon dioxide. Leaving Sundays out, 19 such trains would depart from the power station daily. In contrast, for the high-level waste generated in an entire year by the original nuclear plant—about 45 cubic meters (1,600 cubic feet)—two CASTORs (casks for storage and transport of radioactive material) on a single transport wagon would suffice. Even if low-level and mid-level waste were to be included, amounting to approximately 300 cubic meters (11,000 cubic feet) per year, it would still be quite a manageable portion. It could be carted away by one short freight train.

A last alternative could be to dump liquefied CO_2 in the oceans. But if it is dumped on the upper layers, it will quickly fizz away. So let's pump it into deeper waters. Starting about 500 meters (1,600 feet) below the surface, the pressure is high enough to keep CO_2 liquid and so concentrated that it is heavier than water.[35] If the water is calm enough, the CO_2 will sink ever deeper, forming a CO_2 lake at the bottom of the ocean. But the risk is that deep-sea currents and seaquakes could, over time, cause the gas to be released once again. The IPCC assumes at present that the oceans could serve as CO_2 repositories only for a few hundred years. Furthermore, there is the risk of over-acidification of the oceans, which would have dire consequences for marine flora and fauna. This possibility has not been researched sufficiently to allow dumping carbon dioxide into the oceans to be considered a policy option at this time. For this reason, it is prohibited by the UN Convention on the Law of the Sea, and by an EU directive (enacted in 2009) that makes this ban binding upon EU member states.[36]

Whichever way one looks at it, the final storage problem for fossil-fuel emissions is incommensurably larger than the comparable problem for nuclear power plant waste. This should also point to the fact that the nuclear power option should not be dismissed out of hand.

Space or waste?

Though the disposal of nuclear waste and fossil-fuel waste undoubtedly provides a strong argument in favor of renewable energy sources, the enormous environmental burden imposed by some of these technologies, because of the room they take up, shouldn't be overlooked. Anyone who has illusions about this should fly over northern Germany. White forests of wind turbines stretch as far as the eye can see, their blades milling to produce "green" electricity. True, in 2008 wind power already accounted for 6.4 percent of German electricity production, nine times the energy coming from solar power. However, the most beautiful natural landscapes are being blighted in the name of an apparently environmentally friendly energy source. The problem is again the thin distribution of wind power over the Earth's surface. Though the kinetic power of the wind on our planet would be enough to satisfy all energy demands, collecting this power is Sisyphean work.

A major problem is that the wind is erratic, sometimes still but sometimes so forceful that the turbine blades have to be feathered to prevent them from being damaged. To ensure a sufficiently continuous flow of energy into the grid, it would be necessary to have many more gas-fired power stations standing by to chip in when the wind is still. Usually, hydro power and coal-fired and nuclear power stations are needed to provide a constant flow of electrical power, while gas-fired stations and pumped-storage hydro power plants cover the peak loads. The more electricity is produced from wind, the more often there is a deficiency of supply, which has to be compensated for with electricity from gas-fired power stations, given that the potential sites for pumped-storage plants are limited.

In Europe, electric power is continuously traded via the international grid, with strongly fluctuating intra-day prices resulting from peak load demands and random supply effects. Sometimes, on a Sunday morning when people are still sleeping and the wind is blowing strong, power companies can't shut down their conventional power stations quickly enough and find themselves with an onrush of wind-generated power they don't know what to do with. Since they can't simply get rid of the excess power by, for instance, raising river temperatures to bathtub levels, the power companies, which incidentally are forced to purchase wind-generated power at politically fixed feed-in tariffs, find themselves having to pay external takers to dispose of it. Each year, in Europe, there are a couple of windy "holidays," with negative intra-day prices for electric power.

Thus, to compare wind power with conventional sources of power, one must take account of the randomness of wind supply. Suppose that a country aims for 99 percent supply assuredness, the normal figure when planning electric power capacity. In view of the randomness of wind, only 6 percent of the installed capacity, less than one-third of the actual average power output of wind turbines (which itself is around 19 percent of the installed capacity), can be counted on.[37] However, let us neglect this randomness and assume that the full 19 percent of installed capacity flows continuously and see how many additional wind turbines would have to be erected if their output were to replace that of all EU nuclear plants, which at the moment produce around 935,000 gigawatt-hours per year. The wind turbines in question would be of the most modern, most efficient models, with a 3-megawatt nominal capacity and 0.57 megawatts of actual output. As they generate an average output of about 5 gigawatt-hours per year, one would need 187,000 wind power stations in addition to the 60,000 already in operation in the EU. In other words, the current number of turbines would have to quadruple to 247,000 units. Each of these units measures 100 meters (328 feet) from the ground to the rotor hub, and 150 meters (492 feet) to the uppermost point of the rotor disk. That is 50 meters (55 yards) higher than a nuclear plant's cooling tower.

EU Europe has about 90,000 municipalities. Assuming an average of two churches per municipality, there would be nearly three times as many wind turbines as churches. And each one would be twice as tall as each church spire. Europe indeed would look different.

Wind turbines require a clearance from each other of at least five rotor diameters in the main wind direction, and three diameters at a right angle to it. Thus, one such turbine requires 15 hectares (37 acres). Assuming that we pack wind turbines as tightly as possible, respecting the above clearances, the total area required for the above number of turbines would be 37,050 square kilometers (14,305 square miles)—more than the combined area of Belgium and Luxembourg (or New Jersey and Connecticut).

To make up for a single big nuclear reactor (for example, Germany's biggest, Isar 2, which has a capacity of 1,485 gigawatts and an average output of about 10,000 gigawatt-hours per year), 2,000 such gigantic wind turbines would be needed. They would cover an area 600 times the area inside the fence that surrounded the nuclear plant.

But why stop there? What would happen if, in addition to nuclear power, electric power generated from fossil fuels in the EU were also to be replaced by energy from wind turbines? Nuclear and fossil power together generate 2,910,000 gigawatt-hours in the EU. The wind turbines replacing the fossil-fuel-fired stations would now cover 59,250 square kilometers (22,900 square miles).

Together with the wind turbines now in operation and those needed to compensate for the nuclear power stations, that would be 96,300 square kilometers—an area a bit bigger than Belgium, Luxembourg, the Netherlands, and Slovenia combined, or a little larger than the state of Indiana.

Of course, the wind turbines would have to be placed where there is room available, which in Europe means the countryside. This would mean that there would hardly be any place in Europe where one would not see a wind turbine or hear the hum of its rotor.

The unsightly vista of so many wind turbines could be avoided if we were to banish them to sea, far enough out so that they wouldn't be visible from the coast. Europe's Wadden Sea, along the northern coast of the Netherlands and Germany, one of Europe's last natural spots, is particularly suitable for such a purpose, according to wind power advocates. However, the Wadden Sea covers an area of only 900,000 hectares. To meet the European Union's electric energy needs solely with wind power, in addition to what already is provided by hydropower and other renewable energy sources, would require an area about 11 times as large as the Wadden Sea.

Note again that all this has been calculated under the thoroughly unrealistic assumption that wind blows regularly. With appropriate deductions for its randomness, in order to ensure the 99 percent delivery assurance that is usually required from power plants, the numbers cited above would have to be tripled. But let us stick with the more optimistic assumptions made above and abstract from this randomness.

The figures would be even more impressive if, instead of wind power, solar power were chosen as the alternative. One of the giant, efficient wind turbines assumed above sits on its requisite 15-hectare plot and produces 5.0 gigawatt-hours annually. One hectare (2.47 acres) covered by modern solar panels generates 0.21 gigawatt-hours. Both output figures correspond, by the way, to typical European conditions such as those encountered in Germany. Hence, a 15-hectare plot covered with solar panels would produce 3.15 gigawatt-hours per year, which is 37 percent less than the wind turbine's output. To compensate for its lower energy output, the solar farm would have to cover an area 58.7 percent bigger than that needed by the wind turbine. Thus, Europe would need 56,557 additional square kilometers (21,837 square miles) if all currently existing wind, nuclear, and fossil-fuel power stations were to be replaced with photovoltaic power, disregarding the randomness of supply. An area about the size of Slovakia would have to be added to those countries mentioned above. If we picked EU countries, starting from small to large, as locations to place these solar panels, we would need the first seven of them, namely Malta, Luxembourg, Cyprus, Slovenia, Belgium, the Netherlands, and Denmark, and we would still be 5,000 square kilometers (2,000 square miles) short of the area needed. Picking the smallest US states, we would need Rhode Island, Delaware, Connecticut, New Jersey, New Hampshire, Vermont, Massachusetts, and Hawaii. All of this is obviously nonsense.

If Europe wants to go for the solar option, it would be much better to place the solar panels in southern deserts, where sunshine is much more abundant and where they would not occupy fertile land. After all, south of Europe lies the Sahara, the world's largest desert, with plenty of room where solar panels would not annoy anyone. Utilizing the Sahara for the production of solar energy was an old dream of the influential German engineer Ludwig Bölkow.[38] With the power obtained there, he wanted to produce hydrogen and pipe it to Germany. He chose the conversion to hydrogen because the electric power lines feasible at that time would have been affected by prohibitively high transmission losses.

A variant of this idea is currently been revitalized by a consortium of European power companies and Munich Re (the world's biggest reinsurance company) under the name Desertec.[39] Desertec wants to build a string of power stations in the North African desert, from Morocco to Israel. Instead of transporting the energy in the form of hydrogen, they plan to send it via high-voltage direct-current (HVDC) transmission lines, which have extremely low transmission losses. The 400 planned stations would cover an area of 2,500 square kilometers (965 square miles), about the size of Luxembourg, and would have a capacity of 100 gigawatts, equivalent to 75 nuclear power stations or 140 normal coal-fired power stations. The electricity would be produced by converting water into steam with concentrated sunlight to run conventional turbines, a technique that is being used successfully in

Andalusia. This technology is much more efficient than the photovoltaic panels that currently glitter on European roofs and are beginning to cover farm fields not only in Italy's most beautiful historic landscapes but also in many other places.

A further advantage of the Desertec project is that the solar energy would be stored in containers of high-pressure salt water during the night hours. After sunset, the water from the power station loop would be sent through the heated containers and turned into steam to extend the running time of the turbines. The up-front investment for this project has been calculated at 400 billion euros, or 4,000 euros per kilowatt of installed capacity. Together with the operating costs, this would result in a price of between 6.5 and 16 euro cents per kilowatt-hour—that is, between 30 percent and 200 percent more than the current wholesale price of electricity in Europe, which hovers around 5 euro cents. Unfortunately, the current political situation in the Arab countries doesn't bode well for the Desertec project. Only time will tell how realistic the project is and whether it will turn out to be a breakthrough in the generation of solar power.

How long will the fossil and nuclear fuels last?

Next to slowing down global warming, the advantage of renewable sources of energy would be that they are sustainable, not dependent on the extraction of depletable resources such as fossil carbon and uranium. Since the publication of the Club of Rome's report *The Limits to Growth* in the early 1970s, mankind has been concerned about the possible arrival of "peak oil" (that is, the climax of global petroleum extraction) and the subsequent depletion of fossil-fuel reserves, fearing that the wheels of the modern economy would grind to a standstill.[40] At the time, the forecast was that the oil reserves would be depleted in 40 years. At least this was how the report was understood. Though that was obviously far too pessimistic, it is an undisputed fact that the resource stocks in the Earth's crust are finite and will not last forever.

The stocks of hard coal (anthracite) are huge and are scattered quite evenly around the world. Many countries exploit their own coal deposits and are self-sufficient. The three biggest coal producers are China, the United States, and India, with respective world market shares of 44.9 percent, 17.5 percent, and 8.2 percent, and coal is mined in 60 countries.[41]

Lignite (brown coal) is even more abundant. It is found primarily in the United States (32.2 percent), Russia (31.6 percent), and China (7.3 percent). Surprisingly, however, the world's largest producer is currently Germany, with an 18.4 percent market share, followed by Australia, with 7.4 percent, and Russia, with 7.3 percent. Germany accounts for only 1.7 percent of the global total, but the Swedish firm Vattenfall is currently mining at high speed huge areas in eastern Germany, close to the Polish border, which it was able to buy for a song after the collapse of communism. Lignite is a rather dirty form of coal, with a high sulfur content. The stench of this coal was characteristic to eastern European countries in winter; no visitor could miss it. In addition, it is extracted through opencast mining, and the area isn't always reclaimed afterward. And some European countries don't shy away from sacrificing major historic cultural sites (such as Bohemia) to opencast mining, converting them into moon-like landscapes.

Crude oil is concentrated. The Middle East has 48.8 percent of the proven resources, the Commonwealth of Independent States (a grouping of former Soviet republics) has 16.5 percent, and Africa has 10.4 percent. There are also significant resources in North America (9.4 percent) and South and Central America (7.5 percent).

Most natural gas comes from Russian Siberia, from which a dense net of pipelines leads to Western Europe. Russia holds 36.4 percent of the proven resources, Iran 9.2 percent, and Qatar 6.7 percent. Siberia accounts for 19.4 percent of the world's supply. Next come the United States, with 6.2 percent, and Saudi Arabia with 4.3 percent.

The world's largest uranium deposits are in the United States (18.4 percent), South Africa (10.4 percent), Kazakhstan (10.1 percent), and Russia (9.6 percent). In current extraction, the leader is Canada, with a market share of 23 percent, followed by Australia, with 21 percent, and Kazakhstan, with 16 percent.

It is illustrative to divide the available resource stocks by the current extraction flows, in order to calculate potential running times. Figure 2.6 gives an overview of such running times for the natural resources mentioned, distinguishing among reserves, conventional resources, and unconventional resources. Reserves are defined as the fuel stocks in natural deposits that can be extracted at a cost lower than the current price.[42] Resources are conceptually wider. In this book they are defined so as to include both reserves and proven stocks whose extraction costs are above current prices. The distinction between conventional and unconventional resources is a bit arbitrary. "Unconventional" resources are those made accessible with techniques other than those commonly used today. Examples are tar sands, methane hydrate (frozen natural gas), and reprocessed nuclear waste.

To avoid the mistake of the Club of Rome, we shouldn't construe the running times as predicting when the world will run out of fossil-fuel supplies. They are just simple, and therefore useful, "if-then" statements. If extraction rates remain at today's levels, then the stocks will last until such-and-such a date. This is particularly important as regards running times for reserves. They are definitely not meant to be forecasts, as it is clear that resource prices will rise as depletion comes nearer, which on the one hand will reduce current demand and on the other increase that part of the resources that is classified as reserves. The reserve running time for crude oil, for example, is today about the same 40 years as when *The Limits to Growth* was written, and 40 years from now it could easily still be 40 years. A similar effect could apply to resource running times, insofar as rising prices will trigger more exploration, which would then result in larger known resource stocks. This effect is less important, however, since the days when there were large white spots on the map for sites of natural resources seem to be over. Although detailed exploration is necessary before production can start at a particular site, the general exploration of potential sites has largely been accomplished, notwithstanding occasional news about new sites.

Figure 2.6 shows that, at 41 years, crude oil reserves offer a rather short running time, in the same order of magnitude as uranium, with 40 years, and natural gas, with 59 years. Hard coal and lignite, on the other hand, are so abundant that current extraction rates and prices would guarantee resource owners a profit above extraction costs for another 126 years and another 262 years, respectively.

Unfortunately, looking at conventional resources of crude oil would not add many years to the running times cited above. At rising prices but current extraction rates, conventional oil resources are estimated to be exhausted in only 64 years. Natural gas would last 134 years, and uranium 365 years. Hard coal would last about 2,000 years, lignite about 4,000.

The running times for uranium shown in the figure require some qualifications. Worldwide natural uranium reserves that can be mined with current production technologies at a price of up to $40 per kilogram amount to 1.77 million tons.[43] Around 44,000 tons are extracted yearly.

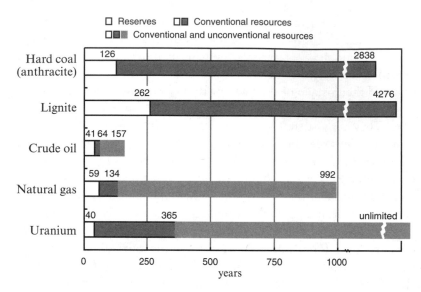

Figure 2.6
Running times. The reserves-to-production ratios are defined as the stocks in 2008 (reserves or resources) divided by the production of each type of fuel in that year. The internationally accepted definition of "resource," which encompasses the reserves, is used here. Sources: Bundesamt für Geowissenschaften und Rohstoffe, *Reserven, Ressourcen und Verfügbarkeit von Energierohstoffen 2009*; Ifo Institute.

This translates into the 40-year running time shown in the figure. However, consumption per year currently is about 64,000 tons. The difference is made up by uranium taken from interim storage, reprocessed fuel rods, and decommissioned nuclear weapons. If the reserves are divided by the rate of current consumption, disregarding the reprocessed fuel rods and the old nuclear weapons, the running time would be only 28 years. Though this suggests that uranium is the least sustainable of the resources considered, the opposite may in fact be true. On the one hand, conventional uranium resources are estimated at 12.67 million metric tons, which gives the mentioned resource running time of 365 years at current extraction rates. On the other hand, the potential for reprocessing nuclear waste, which some countries have abandoned for political reasons even though it is a routine technology, is huge. As was explained above, reprocessing extends the energy produced by a given amount of uranium by a factor of 1.4 or 1.5. That same factor would have to be applied to the running times.

Even more potential lies in fast breeder reactors. These reactors irradiate non-fissile U^{238} to make it capture more neutrons and generate more plutonium than would be the case in a normal reactor. France operates the Phénix breeder reactor for research purposes. Russia, the United States, India, and Japan also operate such reactors. Germany built one at Kalkar but, bowing to political resistance, never activated it. Current uranium prices make breeder reactors uneconomical to operate at this time, but if prices rise this will change. Breeders would extend the energy extracted from natural uranium by a factor of 60. China has just announced that it will build fast breeder reactors, extending the running time of its uranium from 50–70 years to as long as 3,000 years.[44]

Associated with this are the efforts to recover fissile material from spent fuel stored in interim repositories. In the case of "Generation IV" reactors, which operate along the lines of a fast breeder and which probably will become operational in about 20 years, the spent fuel rods could again be used as fuel. Furthermore, such reactors will burn a significant part of the waste directly in the reactor and integrate the reprocessed fuel into the reactor cycle. This would result in much smaller volumes of waste, and the spent fuel would be far less radiotoxic.[45] Generation IV reactors would extend the running times dramatically, and would constitute a very elegant way to reduce the final storage problem.

Another promising line of innovation is being followed in Japan. Japanese researchers have successfully extracted uranium from sea water. Uranium occurs practically everywhere on Earth in low concentrations, even in sea water. Each cubic kilometer (0.24 cubic mile) of ocean water contains about 3.3 metric tons of uranium. The Japanese researchers, using a special membrane, managed to filter this uranium out of the water. The hope is to do this in the near future at an industrial scale that would bring costs down to no more than three times the current price of uranium.[46] Since sea water is available in sufficient volume, and since the cost of natural uranium accounts for only a small portion of the total costs of a nuclear power plant anyway, this appears to be a very attractive option. Large-scale operations are still a long way off, however.

France, Japan, Germany, and other countries also operate a common research reactor called ITER, located in France, that explores nuclear fusion, which is an entirely different method of producing energy from nuclear sources.

Germany also runs its own fusion reactor for research purposes (the Stellarator, in Greifswald, operated by the Max Planck Institute). Unlike fission reactors, which bombard atomic nuclei with neutrons in order to split them, a fusion reactor fuses hydrogen nuclei to form helium, releasing large amounts of energy in the process. In order to achieve fusion, the hydrogen nuclei must be brought very close to each other in order to overcome their mutual electric repulsion—a difficult undertaking, to put it mildly. This requires great pressure and very high temperatures that no known material can withstand. The plasma, a gas made up of the hydrogen isotopes deuterium and tritium in which fusion takes place, has to be confined by strong magnetic fields to keep it from making contact with any solid material. Nuclear fusion generates an extremely low amount of radioactive waste[47] and is exceptionally safe, since any decrease in power levels immediately stops fusion. An explosion or a core meltdown isn't possible. And, the hydrogen input is so tiny that its running time would be, for all practical purposes, infinite.

Cheap and expensive ways to reduce the emission of CO_2

Saving energy and changing the world's energy mix toward "greener" sources are the main ways of fighting the climate problem. In addition, we can try to reduce energy consumption by increasing the efficiency of engines and improving insulation so as to reduce the loss of heat. There are many technical possibilities, ranging from insulating homes to harnessing solar power and bioenergy, and they are all expensive, absorbing resources and manpower that could be used for other purposes. However, some means are less expensive than others, and obviously it is advisable to focus public policy on the least expensive ones. Not only does focusing on the least expensive courses ensure reaching a certain CO_2-abatement goal at minimum cost, and hence with a minimal reduction in society's material standard of living; from a "green" perspective, it also ensures that, given the abatement cost society is willing to bear, a maximum reduction in CO_2 is achieved. Plain economic and ecological considerations come together and adhere to a fundamental efficiency criterion that any rational "green" policy must respect.

Figure 2.7 gives an overview of the abatement costs for alternative measures to reduce CO_2 emissions in Germany, given that country's prevailing wages and weather conditions.[48] Of course, the public subsidies that reduce the private abatement cost below the social cost aren't included in the cost estimates. Though these numbers will be different in other countries and will change over time, the figure nevertheless gives

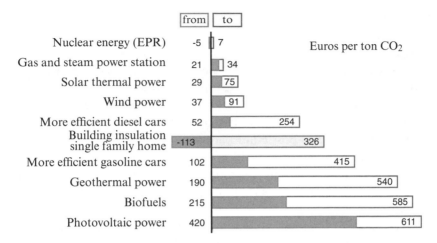

Figure 2.7
Specific costs of CO_2 abatement. The relationships between cost increases and CO_2 reductions relative to alternative reference systems were examined on the basis of concrete projects. For example, in the case of electricity generation, the reference system was a modern anthracite coal-steam power plant burning pulverized coal. In the case of heating, the average German mix for heating of buildings was used as a baseline. Source: U. Fahl, "Optimierter Klimaschutz—CO_2-Vermeidungskosten von Massnahmen im Vergleich," in *Abgas- und Verbrauchsverringerungen—Auswirkungen auf Luftqualität und Treibhauseffekt*, ed. N. Metz and U. Brill (Expert, 2006), pp. 73–94.

a feeling for the magnitudes involved. The alternatives considered range from nuclear energy to photovoltaic cells and include wind energy and measures available to carmakers for new vehicles. The numbers on the left and the right of the vertical line give the range of the abatement costs for alternative technical solutions under the respective category.

Figure 2.7 illustrates how widely the specific costs for reducing CO_2 output vary. This is particularly evident in the case of electricity generation, where the alternatives to a modern coal-fired power station are examined. The cheapest option, at a maximum price of 7 euros to avoid one ton of CO_2 emitted, is to replace a conventional coal-fired station with a nuclear plant. It can even be possible to generate electricity more cheaply with such a plant than with coal, which would imply negative abatement costs—in other words, a gain. The most expensive option is replacing coal-generated electricity with photovoltaic power, the cost of which ranges from 420 to 611 euros per ton. Much cheaper is the replacement of coal-generated electricity with the wind-generated sort.

The avoidance costs for one ton of CO_2 fluctuate in this case between 37 and 91 euros.

After nuclear power and natural gas, the next cheapest option for CO_2 avoidance is solar thermal energy, harnessed by solar collector panels to provide hot water and support heating systems; the abatement costs range from 29 to 75 euros. Close behind are wind turbines as a replacement for fossil-fuel-generated electricity, with costs ranging from 37 to 92 euros. More limited possibilities are offered by replacing older types of vehicle engines with new-fangled, efficient ones. Many carmakers have pushed these boundaries quite far already. Further efficiency gains with modern gasoline engines, at 415 euros per metric ton of CO_2 avoided, are extremely expensive.

A similar case is that of building insulation. With an old house, more effective insulation can lead to substantial cash savings, at a cost of up to 113 euros per metric ton of CO_2 avoided. However, further improving the insulation of a new house may cost as much as 326 euros to curtail an additional metric ton of CO_2.

The three last rungs are occupied by geothermal power, biofuels, and photovoltaic cells on roofs. Geothermal energy and solar cells are extremely expensive. Biofuels avoid little CO_2 output, if any. Should the Desertec project ever become reality, the abatement cost for solar power could be significantly lower than the cost given in figure 2.7. However, in the foreseeable future solar power is not expected to become competitive with wind energy or solar thermal power, let alone nuclear power.

The law of one price and the European emissions trading system

Though at a specified point in time it might be possible to find out what the cheapest abatement technologies are, technological progress is not foreseeable. For each of the categories depicted in figure 2.7, there are dozens if not hundreds of subcategories of further variants of the respective technologies, and their ranking is subject to permanent change due to technological progress and changes in prices of the respective raw materials needed, as well as changes in wages. And even the ranking of the broad technology categories depicted above may change. Who knows, perhaps one day solar energy will indeed turn out to be the cheapest of all. How can one make sure that society chooses the most efficient ways of curtailing CO_2 emissions under such conditions?

Many may think that this should be decided by government bureaucrats or through democratic debate in parliaments. However, this would be a central-planning approach, and it would be doomed to certain failure because the political bodies would not be able to gather the necessary information and would not have the incentives to choose the right technologies. Rather, it is likely that the incumbent producers of "green" technologies—those who happened to get in on the ground floor—would be able to influence the political decision process so as to tilt the policy decisions in their own direction. This is obvious when one compares France and Germany. Whereas in France nuclear energy is thought of as a *patrimoine nationale* (national treasure), German politicians abhor it and praise wind and solar power instead. France is a world champion in the production of nuclear-generated electricity; Germany is a champion in wind and solar energy. In each country, the vested interest of the respective industries, with hundreds of thousands of jobs at stake, has succeeded in swinging public opinion toward a particular idiosyncratic ideological predisposition.

Economists, in contrast with politicians, insist on having the market select the cheapest technologies through its "law of one price." Both an emissions tax and an emissions trading system establishing a common price per ton of CO_2 emitted would provide the proper incentives and would ensure that market agents would select the cheapest abatement technologies. If everyone has to pay a certain price per ton of CO_2 emitted, they will all carry out those abatement activities available to them that, per ton of abatement, are cheaper than this price. It is useful to imagine that the polluters order their respective abatement tools inversely to the abatement cost per ton, so that, for each polluter, marginal abatement costs per ton rise with an increase in the total quantity this polluter abates. The polluters will then consider moving through the ranked set of abatement technologies until they reach a point where they are indifferent at the margin between, on the one hand, emitting another ton and paying the emission price and, on the other, abating the ton and footing the abatement cost. At that point all cheaper means to abate at a cost below the emissions price would have been exhausted, and all the more expensive means would remain unrealized. As all polluters face the same emission price, this allocation of abatement activities across all firms is socially efficient, even though there is no one who possesses the collective information about the available abatement technologies and no one who dictates the allocation of abatement activities to the single firms. Based on the law of one price, the economy partially regulates itself as if steered by an invisible hand.

Of course, the self-regulation is only partial. A central authority is necessary to fix the emissions tax rate or to determine the overall number of emission rights in a trading system, so that the market can find the appropriate price. These functions can't be provided by the market, since environmental quality, in terms of excessive CO_2 being absent from the atmosphere, is a public good, benefiting all rather than only those who provide it. Nevertheless, polluting firms make the socially correct secondary decisions given these primary decisions by policy makers, even though no firm is driven by altruistic motives and no firm reveals its abatement options.

Skeptics may counter that they could ask the firms to reveal their abatement options and then command implementation of the best abatement technologies. But then every firm would have a strategic interest in overstating its true abatement cost, in the hope that this would result in the imposition of correspondingly low abatement standards for itself. Only the implicit self-revelation of marginal abatement costs under the law of one price, driven by the profit motive, leads to unbiased and socially efficient abatement strategies. The law of one price makes it possible to minimize the aggregate abatement cost for any particular quality of the environment, and, given the aggregate abatement cost society is willing to accept, it maximizes the quality of the environment.

The European Union has taken this insight to heart in its environmental policy, with a view to fulfilling its Kyoto Protocol commitments. In October 2003, it issued a binding directive to its member countries establishing an emissions trading system for CO_2 and other greenhouse gases.[49] That directive came into force in 2005. Such a system is sometimes also called a *cap-and-trade* system. Its essence is that the regulator determines the total amount of emissions available (the cap) by issuing a limited number of tradable rights or emission certificates. The cap is specified for a well-defined trading period, and the rights expire when that period ends. The certificates are known as European Union Allowances (EUAs). They consist of virtual units that are carried in an EU accounting system. The trading system covers all 27 EU member states.

For the time being, the EU system gives the certificates to the companies at no charge, in an endowment, on the basis of a politically determined allocation plan. The companies can expand or decrease their endowments by trading the certificates in a common market organized like a stock market, in the process generating a market price for CO_2 emissions. If a company wants to emit more than its original endowment implies, it must buy additional allowances; if it emits less than its endowment, it can then sell the unneeded allowance certificates.

Being able to sell the certificates doesn't interfere with the allocation function exerted by the law of one price; much to the contrary, in fact. As was explained above, the company that has to buy the certificates extends its abatement activity to the point where its marginal abatement cost equals the price of the certificate or emission right. The same rule is followed by the selling company. It realizes all abatement possibilities that are cheaper per ton than the price it could collect for a certificate, and hence the marginal abatement tool it employs is one that costs just as much per ton as the company can earn by selling the certificate set free by this tool. Going further and abating greater quantities at a cost above the certificate price would obviously not be worthwhile, because the revenue from selling the extra certificates would fall short of the extra abatement cost. Thus, the selling company and the buying company follow exactly the same socially efficient abatement rule, as the Coase Theorem[50] would predict.

Trading can be conducted bilaterally between companies (over the counter) or, more formally, through brokers and on special markets. The biggest of the markets—the European Energy Exchange (EEX), located in the German city of Leipzig—is used by more than 200 trading entities from 20 countries.

A fundamental prerequisite of an emissions trading system is a carbon registry. In Europe each country runs one such national registry, keeping allowance accounts for all its companies and supervised by a national authority; the national registries are consolidated in a European clearing house in Brussels. The national supervisory authorities also control the CO_2 emissions by overseeing an accounting system for the inputs of fossil fuels and calculating the emissions on the basis of technological coefficients. Under no circumstances may a company emit more, within a well-defined trading period, than is covered by the allowances it possesses. If the company emits CO_2 beyond its allowances, it must pay a fine for the tons of CO_2 emitted in excess. During the first trading period, this fine was set at 40 euros per metric ton—well above the allowance certificate price, which mostly ranged from 15 to 25 euros. In addition, the excess tons would be deducted from the company's endowment in the next trading period. The fines are collected with draconian zeal, and the emissions are controlled very closely. No major breaches of contract in allowance trading have been reported.

The EU trading system encompasses the energy producers and the energy-intensive industries, including power generators, oil refineries, coking plants, iron and steel works, and the cement, glass, chalk, brick, ceramics, cellulose, and paper industries, but initially not the chemical industry. The crackers in the chemical industry that produce light oil from heavy oils were, however, included in 2008 because their high CO_2 output.

The trading system covers 45 percent of the European Union's CO_2 emissions, and about 30 percent of its greenhouse-gas emissions if the other greenhouse gases are taken into account.[51] In Germany, the trading system encompasses as much as 51 percent of carbon dioxide emissions.[52]

The power utilities account for the largest share of the economic sector affected by the trading of emission rights, 32 percent of the CO_2 emissions covered in Europe being integrated practically without exception in the trading system. The only ones left out are the very small power stations with a combustion capacity below 20 megawatts or a power-generating capacity below 7 megawatts (about a hundredth of the capacity of a normal coal-fired power plant). All small power plants put together account for only 1 percent of the electricity generated. Thus, the European emissions trading system covers 99 percent of all electric power generated in the EU. The practically all-embracing inclusion of the power stations turns out to be important for passing judgment on the national energy policies.

Emissions from the other sectors, in particular fuel combustion for heating of homes and transportation, but also most of the manufacturing industries, aren't integrated into the trading system, even though they form part of the Kyoto Protocol. The system is to be extended gradually in the coming years to the remaining sectors of the economy. Starting in 2011, intra-European flights will be brought into the trading system, and beginning in 2012 international flights too will be brought in.[53]

The European Union specified two legally binding trading periods for the exchange of allowances and submitted proposals for a third. The first period (2005–2007) had a test-phase character. The second phase goes from 2008 until 2012, and the third phase will go from 2013 to 2020.[54] The allowances may not be brought forward from one period to the next, as they expire at the end of the respective period. For the first period, an overall amount of 2.19 gigatons of CO_2 was permitted. That was 11.7 percent less than the 1990 emissions, which amounted to 2.48 gigatons, but 14 percent more than the average of the years 2000–2002, which had been taken as the baseline for allowance trading, providing an overly generous surplus for economic growth.[55] When trading started, the companies offered prices of nearly 30 euros per ton until the spring of 2006.[56] But the price then nosedived, stabilized for a couple of months, and subsequently fell to near zero, where it stayed from the spring of 2007 until the end of that year.[57] The price decline presumably resulted from the fact that companies hoarded certificates as a buffer for uncertain future needs, which naturally lost their value as the time of expiry approached. Also, the European Union may have been too generous in the first trading period, wishing to forestall political resistance that would have jeopardized the whole project.

The second trading period has started, and it will last until 2012. For this period, only 2.081 gigatons of CO_2 are allowed—5.0 percent less than in the first period.[58] This is still 8.4 percent more than during the reference years 2000–2002, but not as much as would have been the case with normal economic growth and no particular efforts to save energy. The scantier allocation pushed up the allowance price by mid 2008 back to a level around 27 euros per metric ton of CO_2, close to the price it had reached by the middle of the previous trading period. However, during 2010 the price hovered at around 14.30 euros, probably again because the end of the trading period was approaching.

In order to avoid a precipitous decrease between trading periods, the trading system will be fundamentally different from 2013 on. For one thing, the allowances are to be allocated (or auctioned) on a yearly basis; for another, they will be valid over a longer period, albeit subject to depreciation by a certain percentage from year to year.[59] The depreciation and the yearly allocation of new allowances will enable the European Commission to steer and stabilize more finely the evolution of emissions and allowance prices.

In the future, the European Commission wants to auction the allowances instead of allocating them without charge.[60] As early as the beginning of the third trading period, in 2013, the entire amount of allowances for the electricity branch, which accounts for the largest share, will be auctioned, the revenue going to the respective national governments. The German government already counts on auction revenues from 2013 of up to 10 billion euros yearly.[61] Starting in 2020, all allowances are to be auctioned, and the remaining sectors of the economy are to be brought gradually into the system.

When the trading rights become more scarce, this may turn into a fairly expensive affair for companies in need of them. In the end, the air belongs to the general public, represented by the state, and the companies must pay the full amount for using it, just as they do for their other raw materials. The internal logic of this arrangement can't be faulted. The environment is yet another production factor, and if someone wants to use it, its owner must receive payment in full.

The trading system for European Union Allowances makes the EU a sort of guinea pig for the world, since Article 17 of the Kyoto Protocol of 1997 provides for more such systems to be established among the signatory countries. Moreover, the EU system was the model for an international trading system that the United Nations introduced in 2008 for the 52 countries that have accepted binding emission constraints, as was explained above. The UN christened its certificates Assigned Amount Units (AAUs). The European Union Allowances and the UN units are essentially the same and represent, as it were, the same currency. They allow the emission of one metric ton of CO_2 per trading period and are listed in an integrated UN-EU accounting system. They are as intimately related as the one-euro coins that feature differing national symbols on their flip sides but all represent the same value. One important difference, however, is that the UN units are tradable between states, not between individual companies. If a country has an excess of AAUs, it can sell them on to another country, and the buying country can decide, through legal measures applying in its jurisdiction, which of its own industries will be allowed to emit more CO_2. Conversely, a country that is having difficulty meeting its reduction targets can buy the necessary permits from other countries. This represents a decisive breakthrough toward an efficient control and allocation of emissions among the Kyoto countries, but unfortunately only 30 percent of worldwide emissions are currently captured by the system.

The UN assumes that in the future other groups of countries will adopt an allowance trading system similar to that of the EU, or that the EU will expand its system to include other countries. One idea that is being floated is to integrate California into the EU system; that may sound a bit odd but it is fully compatible with the global nature of the climate problem.[62]

Feed-in tariffs, instrumental goals, and European policy chaos

The law of one price is the law of economics, the secret of the market economy's success, and the foundation for the advantage that the emissions trading system provides for a rational "green" policy. However, it conflicts sharply with the law of politics. A rational CO_2 policy would define the abatement goal and then leave it to specialists to find the best instruments to achieve it. The political value judgment would apply to the goal, but not to the instruments. Real policy, regrettably, doesn't work that way, as it doesn't distinguish between goals and instruments, instead turning instruments into goals by attaching value judgments to them. Thus, in practice, policy is determined by a chaotic blend of goals and instruments that follows no other logic than the stubborn preferences of the political process.

The CO_2 policy of the European Union and its member countries is full of examples of this. One is the 20–20–20 goal that the EU Commission formulated in a directive issued in 2007.[63] According to this goal, the EU countries must reduce their CO_2 emissions by 20 percent on average relative to 1990, and they must increase to 20 percent on average the share of renewable energy in total energy consumption by January 2020, both goals being broken down to specific targets that differ from country to country. The directive also stipulates that energy use for heating at a given room temperature is to be brought down by 20 percent. Obviously, this directive mixes goals and instruments, because increasing the share of renewable energy is an instrument that would help achieve the goal of reducing emissions. With an emissions trading system in operation that covers 99 percent of the electricity production and ensures, because of the law of one price, that the emissions reduction is allocated to the EU's power plants in such a way as to minimize the overall abatement cost, the extra goal of reaching a renewables share of 20 percent probably will entail unnecessary additional abatement costs. Only the fact that the 20 percent target for renewables can also be reached in areas of the economy not covered by emissions trading, such as heating and transportation, provides an extra degree of freedom to meet the climate goal and the instrumental goal separately. But this is theory.

In practice, most EU countries make huge efforts to promote "green" electricity to meet the respective national targets for the share of renewable energy that the EU has set them. In the 27 EU countries, there are hundreds, if not thousands, of national decrees, laws, and subsidy programs for "green" electricity, at various levels of government, from the federal government down to local municipalities that aim at promoting electricity from renewable sources. As long as these policy measures are operative, they distort the efforts to reduce CO_2 emissions brought about by the emissions trading system, and hence increase abatement costs, because a suboptimal mix of power generators is then installed.

The policy of promoting "green" electricity by way of feed-in tariffs, very popular in Europe, is a particularly important example of this type of policy irrationality. Feed-in tariffs remunerate electric power delivered to the grid at prices above the wholesale market prices for electricity, prices high enough to cover the costs of producing "green" electricity with hydro, wind, or solar power, combined with a purchase obligation for the utility running the grid. They are guaranteed for a long period of time so as to eliminate the investment risk for the supplier. The grid operator is allowed to raise the retail price of the electricity it sells to final consumers to compensate for his extra cost. As the increase in the retail price doesn't discriminate between the sources of the electricity, consumers have neither the incentive nor the possibility to avoid the expensive electricity from renewables. Although feed-in tariffs operate outside the public budget, they effectively operate like subsidies paid to "green" electricity producers, financed with a general tax on electricity consumption. The 20 EU countries that use such feed-in tariffs are Austria, Bulgaria, Cyprus, the Czech Republic, Denmark, Estonia, France, Germany, Greece, Hungary, Ireland, Italy, Latvia, Lithuania, Luxembourg, the Netherlands, Portugal, Slovakia, Slovenia, and Spain.

These countries have set prices for solar electricity ranging from 30 percent (Slovenia) to 1,020 percent (Luxembourg) above the normal wholesale base load electricity price (currently 5 euro cents). The feed-in tariffs for wind and hydro power are usually lower, as these technologies are closer to the break-even point. However, price surcharges of between 80 percent (Germany) and 160 percent (Latvia) are not uncommon. The costs to consumers are gigantic. In Germany alone, the subsidies guaranteed for the photovoltaic devices already installed amount to approximately 85 billion euros, despite the fact that the increasingly nervous German government hastily decided in 2010 to reduce the feed-in tariffs for new stations.[64]

The most obvious shortcoming of the feed-in tariffs is that they can't affect Europe's overall volume of CO_2 emissions, because it has been fixed by the cap that the European Union settled upon—that is, by the number of emission certificates it has issued.[65] It is possible, of course, for a single country to emit less CO_2 if it offers sufficiently high feed-in tariffs, because transportation costs for electricity will imply that the "green" electricity these tariffs promote will crowd out conventional electricity generated from fossil sources in the neighborhood of the "green" power plants. However, this also leads to emission certificates becoming redundant and being offered for sale, or alternatively to their not being bought at the European exchange market, enabling more distant fossil-fuel-powered stations to emit exactly as much additional CO_2 as has been curtailed in the neighborhood of the "green" plants. With its feed-in tariffs, the "green" country reduces the market price of emission certificates and effectively subsidizes the fossil-fuel power plants elsewhere in Europe. It is unbelievable, but true, that the politicians of the countries with feed-in tariffs simply overlook the fact that these tariffs are entirely ineffective because they conflict with the European emissions trading system.

Worse, the feed-in tariffs will not even promote "green" technologies, but will instead impede their development in other countries if those other countries don't counter them with their own feed-in tariffs. The reason is that the fall in the price of emission certificates they cause reduces both the cost of producing electricity from fossil sources and the wholesale price of electric power, making it more difficult for providers of renewable energy to establish themselves in the market. The more solar power that feed-in tariffs promote in Denmark, where sunshine is a rare commodity, the more solar power will be crowded out in Spain's Extremadura, where sunshine is abundant. Every wind turbine erected on Danish fields and every additional solar panel glittering on a German roof as a result of the feed-in tariff will increase greenhouse-gas output in the rest of Europe as much as it reduces it in Germany or Denmark. The new Spanish solar power complex will not be built, Italians will continue to use conventionally generated electricity for their air conditioning, and Poles will refrain from modernizing their coal-fired power stations. Even the French might give up their production plan for an additional nuclear plant, investing instead in a gas-fired power station.

The price of emissions certificates already stimulates the production of "green" electricity, without resorting to feed-in tariffs. And because it respects the law of one price, it does so in an efficient manner. Across

Europe, the market picks the most efficient locations for "green" electricity to be generated at the lowest cost. Feed-in tariffs, in contrast, relocate "green" power stations from the countries that possess a natural advantage to those that offer feed-in tariffs, while the amount of fossil fuels burned remains unchanged. As a result, electricity generated from renewable sources will be produced at a higher cost than is necessary.

Even research is distorted, as solar panels and wind power stations will not be optimized for operating in the best natural locations, and instead will be optimized for the countries offering feed-in tariffs. Those locations usually don't coincide.

It is sometimes argued that, although feed-in tariffs may not affect CO_2 output, if the cap is fixed, they will enable a further tightening of the cap in the next trading period, because they help establish an industry for renewable power technologies that makes it easier to set more demanding abatement goals.[66] This argument, however, is thoroughly wrong. Since feed-in tariffs distort the renewables market across Europe and increase the cost of "green" electricity, which consumers end up paying for, relative to what the emissions trading market would have accomplished alone, they will definitely increase rather than reduce public resistance against a further tightening of the cap in the next trading period. The larger the pain associated with the CO_2-abatement policy, the higher the resistance against it.

But it could be argued that the promotion of renewable energy, even if it does nothing for the climate, should at least help the countries promoting it to meet their Kyoto commitments. After all, most European countries have entered into a binding commitment to reduce their CO_2 emissions by a substantial percentage over the years 2008–2012. If the entire reduction in national CO_2 emissions resulting from wind turbines and solar-paneled roofs made possible by the availability of feed-in tariffs were credited toward the country's Kyoto goals, that country would be able to satisfy its commitments, even if the allowances it no longer needs were sold abroad, thus enabling the buying countries' CO_2 emissions to rise to higher levels.

Unfortunately, that isn't the case either, since exported emission rights in the Kyoto account are treated *as if* the corresponding emission had taken place in the exporting country. If a country's power stations buy more emission certificates in order to be able to emit more CO_2, the country's formal emission goals under the Kyoto Protocol will not be jeopardized, because the additional CO_2 emissions made possible by acquiring such certificates will be charged to the account of the selling country. Conversely, a country will not come nearer its Kyoto reduction goals if it sells certificates, since the emissions of the buying country will be tallied to its own account. Even though a country may be very successful in stimulating the production of electricity from renewable sources and crowding out its fossil-fuel power stations, it will not get any closer to meeting its Kyoto commitments, because it is not the actual emissions that count, but only the emission certificates it was allocated.

It is true that by trading AAUs (the emission rights traded under the auspices of the UN) a country can alter its Kyoto Protocol commitment. It can relax the commitment through the acquisition of AAUs, or tighten it through their sale. However, shifting the actual emissions, as is the case between companies through allowance trading in Europe (with EUAs), doesn't translate into shifting the AAUs between countries. Such shifts aren't taken into account for the Kyoto balance. If, for example, a German power plant sells part of its allowances to Poland, because it switched from fossil fuels to "green" fuels and thus reduced its emissions, the additional Polish emissions will count in the UN system as if they had been emitted in Germany. This will be important when this trading phase expires at the end of 2012 and a review is conducted of each country's compliance with its Kyoto Protocol commitments. For this purpose, a country's emissions occurring outside the sectors covered by the trading system will be added to the emissions permits it was initially allocated, and this sum will then be compared against the reduction target set originally. How much a certain country actually emitted in the sectors covered by the permit trading plays no role for the Kyoto balance, as deviations from the initial allocation of permits will always be offset by equal and opposite deviations in other European countries. Of course, the preconditions are that all companies play by the rules and that none emits more than what the trading rights it possesses permit (a practice the EU controls very strictly, imposing hefty fines on violators).

These strange accounting methods may be one reason why Denmark is far from fulfilling its commitment. Although the Danes had promised to reduce their emissions by 21 percent from 1990 to the target period 2008–2012, they had decreased them only by 4.0 percent by 2008, which makes it nearly impossible for them to come close to the target in the time remaining. That appears puzzling at first sight, because Denmark evidently makes great efforts to generate its electricity with wind turbines—it is even more densely covered by wind turbines than Germany. Whereas Germany managed to generate 6.4 percent of its electricity from wind power in 2008, Denmark had already achieved 18.4 percent. Although Denmark has only 5.5 million inhabitants, it is already the world's sixth-largest wind power producer.

But the puzzle ceases to be a puzzle when one considers how the Kyoto balance sheet is put together. Denmark can erect as many wind turbines as it pleases, but it still will not come one bit closer to fulfilling its reduction commitments, since the emissions certificates allocated to it, which then are set free and sold to other countries, count as if the emissions had taken place in Denmark, whether or not they actually did. A similar observation can be made with regard to Spain, the world's second-largest wind power producer, which also has dismally failed to meet its Kyoto commitments. While Spain had been allowed to increase its emissions by 15 percent, its actual emissions increased by 54.3 percent by 2008. Wind power crowding out fossil power doesn't translate into emissions being crowded out of Spain's Kyoto account.

These disturbing considerations show that the double intervention of "green" power policies and trading of emissions rights doesn't make sense. It would be better for all European countries, and neutral for the climate, if those countries would terminate their feed-in tariffs and rely entirely on the incentive mechanisms of the emissions trading scheme.

3

Table or Tank?

Capturing the sun

Civilization developed with the help of bioenergy. When our ancestors enjoyed the heat of a fire, or when they used animal muscle power for transportation, the source of energy was plants. In many regions of the world, plants still are the dominant source of energy today.

Bioenergy is nothing more than stored sunlight. In photosynthesis, plants use sunlight to transform carbon dioxide and water into carbohydrates, stripping the oxygen from the carbon and latching hydrogen to it instead. These carbohydrates contain what chemists call *reduced carbon*. Out of them a plant makes the complex biological compounds, such as sugars, starch, and cellulose, that constitute both the plant's structure and its fruit.

Reduced carbon can be burned in order to retrieve the energy it contains. This can be accomplished, for example, by burning wood to make a bonfire, or by eating plants and burning in the body the carbon compounds, fats, proteins, and carbohydrates they contain. When burned, carbon oxidizes, that is, binds again with oxygen. Oxidation and burning are one and the same; whether a flame occurs is immaterial. Photosynthesis uses energy to break down carbon dioxide into carbon and oxygen. Burning releases this energy once again, as carbon and oxygen bind to each other.[1] When storing energy, the same amount of CO_2 will be captured as will be released through combustion. For this reason, using biomass as a fuel source is, under ideal conditions, CO_2 neutral.[2]

In nature, photosynthesis and plant decay are in equilibrium. Plants grow, die, and decay. When they decay in the presence of oxygen, the carbon contained in them burns, and carbon dioxide is released. If this occurs in the absence of oxygen, methane will be released. But, as was explained in chapter 1, methane oxidizes to carbon dioxide over an average of 15 years when it comes into contact with the oxygen in the atmosphere. Photosynthesis then again strips the oxygen from carbon dioxide, generating reduced carbon and thus creating biomass. This closes the circle.

Carbon dioxide is also absorbed by the oceans, loosely binding itself with the water in the upper layers to form carbonic acid; then it is eventually released back into the atmosphere through the action of waves. The saturation of the upper layers of the world's oceans with CO_2 is closely related, at a given temperature, with the CO_2 saturation of the atmosphere. In a closed cycle involving the air, biomass, and the upper layers of the oceans, a nearly fixed amount of CO_2 circulates over very long periods. At any given time, the carbon is bound in certain proportion in the oceans' upper layers, in the atmosphere, and in the biomass.

Mankind lived as part of the biological carbon cycle until about the middle of the eighteenth century, when industrialization began around the coal mines of Sheffield in England. With the invention of the steam engine and all the other combustion engines that followed, the wheels of industrialization began to spin, leading to rapid economic growth that gave rise to the standard of living that Western civilization enjoys and to which Asian societies now aspire. However, mankind is beginning to realize that the carbon exhaust that industrialization has released into the atmosphere is a serious debt left to future generations. A substantial fraction of the fossil carbon that is added to the cycle, about a quarter in the long-run, accumulates in the atmosphere and warms the planet, as CO_2 hinders the incoming visible short-wave radiation from the sun being reflected back to space as invisible long-wave thermal radiation. This is the rationale for returning to biological energy sources. If biomass is used as an energy carrier, no additional CO_2 will be added to the carbon cycle, and hence the saturation of the atmosphere with CO_2 will not increase.

Of course this benefit is also offered by wind, solar, hydro, and nuclear power, to name the most important technological options. However, bioenergy may possess the greatest beneficial potential, as plants are a natural and cheap device for capturing thinly dispersed energy. As was explained previously, renewable energy, though abundant, is thinly dispersed across the planet's surface and therefore can be used as a replacement for fossil fuels only if techniques are found to concentrate it in particular locations at reasonable cost. Modern planting and harvesting methods do make use of such techniques.

However, a number of conditions must be met for bioenergy to be really climate neutral and to ensure that it doesn't add more carbon dioxide to the atmosphere. The most important of these conditions is that we use only biomass that otherwise would have been left to decay, and that we don't burn this biomass in order to extract its energy any earlier than it would have decayed naturally. If we burn wood, for example, we accelerate the decay process by releasing into the atmosphere CO_2 that otherwise would have remained trapped in the wood longer, thus altering the apportionment of CO_2 among the three storage media described above. The newly released CO_2 will stay in the atmosphere and contribute to the greenhouse effect. Only after many decades will it be absorbed by the oceans, and ultimately by freshly growing biomass, until the old equilibrium is attained once again. It is estimated that the Earth would take about 300 years on average to regain its old balance. It might take longer, depending on the type of tree doing the absorption. If the giant sequoias, which live and grow for thousands of years, were to be felled, this would be akin to extracting fossil fuels, as it would dramatically diminish the stock of reduced carbon on Earth. Fortunately, the US Supreme Court ruled that the George W. Bush administration's plan to allow just that was illegal.[3] Seen this way, logging to use wood as fuel is not climate neutral.

The best we can do as regards biomass and climate is to help increase the biomass stock instead of burning it. Afforestation is, for this reason, an important contributor to slowing down climate change. The bigger the stock of forests on our planet, the more CO_2 is bound in biomass, instead of doing damage in the atmosphere.

For the same reason, using biomass to build houses and to make furniture poses no problem for the climate; indeed it is beneficial. Wood used for these purposes is kept dry and protected from oxidation, at least until the house or the furniture is disposed of. Construction timber not only stores carbon; it also makes room in the forest for new trees, so that the overall carbon stock stored in biomass increases, while that in the atmosphere decreases. From this point of view, both the construction industry and the furniture industry should be encouraged, as they are beneficial for our climate. And perhaps the social disapproval of the use of teakwood should be withdrawn, although in that case the situation is more complex. (In order to set up a teakwood plantation, it is often necessary to clear tracts of jungle.)

Burning biomass to tap its energy is climate neutral only if it doesn't change the stock of living biomass. In practice, a policy of planting energy-rich plants such as corn (maize) or sugarcane on fallow fields comes close to this condition. The energy obtained in this way would entail only a minor reduction in the planet's biomass stock, being therefore largely CO_2 neutral. And if the bioenergy obtained that way replaces fossil-fuel energy, it will even slow down climate change.[4] But even when such is the case (something that, for reasons that will be discussed in chapters 4 and 5, can hardly be expected), the condition must be met that during cultivation and fertilizing no greenhouse gases are released. Unfortunately, that condition is seldom met.[5]

Competition between bioenergy crops and food crops must also be considered. It would be ideal if only fallow land were used for energy crops. In practice, unfortunately, that is seldom so. In most cases, energy crops are raised on land that would otherwise be used for food crops. This pushes up the prices of foodstuffs, and it can lead to shortages in poorer countries.

What is bioenergy?

Bioenergy is the most important sort of renewable energy, far outperforming hydro, solar, and wind power. In fact, it is a more important source of renewable energy than all the other sources taken together, accounting for 55 percent of the consumption of renewable energy in the OECD countries. The worldwide figure is even higher: 79 percent. In the United States it accounts for 65 percent, and in the European Union for 68 percent.

Humans have obtained energy by burning wood for millennia, and wood continues to be the most important source of bioenergy to this day. As figure 3.1 shows, in the OECD countries the burning of wood accounts for 72 percent of the bioenergy used. The worldwide share is even higher, at 94 percent, as developing and emerging countries are even more reliant on this energy source than developed countries.

The burning of wood to tap its energy need not be done in the traditional way—that is, by combusting logs in a hearth or a furnace. In Europe, compacted wood pellets have become very popular. The Netherlands and Belgium even subsidize their use. Pellets are nearly as easy

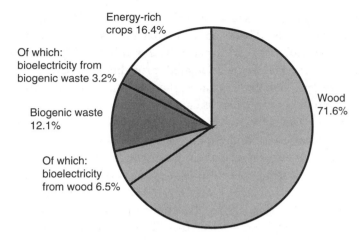

Figure 3.1
Bioenergy in OECD member countries, 2007. Sources: International Energy Agency, *World Energy Outlook 2009*, *Renewables Information 2009*, and *Electricity Information 2009*; estimates and calculations by Ifo Institute.

to transport as liquid fuels. They are delivered by tank trucks and pumped into special containers in the cellars of homes, from which they are automatically fed to the home's furnace or boiler. Heating with pellets has an environmentally friendly aura and is often cheaper than heating with oil. In Germany, a liter of heating oil, including the value-added tax, cost 0.59 euro in 2007; the equivalent amount of pellets (2.3 kilograms) cost only 0.47 euro.[6] However, the truth is a bit different.

For one thing, burning pellets produces more residue than burning oil or gas. The particulate matter contained in the smoke is unhealthy and has an unpleasant smell, a bit like the awful lignite furnaces used in communist times throughout eastern Europe. If a homeowner wants to be "greener" and buys a pellet-fueled heating system to protect the environment, he or she should consider that the quality of the air in the neighborhood will deteriorate substantially as a result, which may bring down the prices of the affected homes.

For another, the production of pellets can be a rather dirty process. Pellets once were made from sawdust and wood shavings from sawmills. But European demand is now so huge that pellets are being increasingly imported from Eastern Europe—mostly from Belarus, where firewood is ground and then dried and heated with the country's cheap natural gas until its lignin becomes liquefied, which makes it possible to compress the wood into pellets. Lignin hardens again when it cools off, helping the pellets retain their form. Their shape is what makes it possible to pump them automatically into a furnace. The eco-balance of these imported pellets, which hasn't been calculated, probably is appalling.

A relative of pellets, one that reduces the environmental damage caused during the production process, is chopped wood. Chopped wood has become more popular recently because industry has provided inexpensive automatic furnaces for home use.

A more sophisticated way to get energy from wood is to use a wood gasifier to produce *synthesis gas*. Synthesis gas is a mixture of carbon monoxide and hydrogen produced by smoldering hydrocarbons at a low oxygen level. The synthesis gas is typically used to power a gas engine which turns a turbine to generate electricity. This form of electricity accounts for 6.5 percentage points of the share of wood in bioenergy, which is equivalent to 80 percent of the electricity produced by all wind power stations.

After wood, energy crops and biogenic waste are the second and third most important sources of bioenergy. Energy crops include sugarcane, oil palms, corn, rice, wheat, rye, rape, sugar beets, sunflowers, soybeans, and potatoes. Ethanol, biodiesel fuel, and methane can be produced from their fruit, seeds, and tubers. In the OECD countries, energy crops account for 16.4 percent of bioenergy; globally their share is about 4 percent. The United States, Brazil, Indonesia, and Germany are particularly active in the cultivation of energy crops. In the US, energy from these crops accounts for nearly 19 percent of all renewable energy, far above the 7 percent EU average. Germany, however, stands out in the EU, with a 19 percent share.[7]

Biogenic waste includes straw, wood residues, slurry, landfill gas, manure, sewage sludge, logging residues, slaughterhouse waste, and household waste, to name only the most important sorts. It can be fermented to obtain methane, or it can be burned directly. Methane from fermentation is used to power gas engines, sometimes in combined heat-and-power-generation plants that heat farmhouses and produce electricity as a by-product. Of all the uses of bioenergy this is probably the most promising, as it exploits, with a well-established and mature technology, energy that otherwise would have been lost through natural decay and fermentation. Biogenic waste contributes 12.1 percent to bioenergy in the OECD countries, 3.2 percentage points of which go into electricity production.

Additionally, efforts are being made to produce liquid fuels from biogenic waste through complex chemical processes.

"Green" gasoline

The most exciting and significant prospects for bioenergy are offered by the production of biofuels, that is, bioethanol and biodiesel fuel. Currently bioethanol is far ahead, with 76 billion liters (20.1 billion gallons) produced in 2009 from alcohol distilled from biomass. Biodiesel fuel amounted only to 12.6 billion liters (3.3 billion gallons).[8]

Bioethanol can be blended with gasoline at a concentration as high as 20 percent without damaging conventional engines, and specially designed engines can even run on pure ethanol. However, ethanol has a lower energy content than gasoline, requiring a volume 54 percent greater to yield the same performance. Because ethanol is alcohol, a bile-like substance is added to make it unpalatable. Like hard liquor, bioethanol is produced by adding yeast to starchy feedstock, letting it ferment, and then distilling the resulting alcohol.

The trailblazers in developing bioethanol were the United States and Brazil. In the US, the usual feedstock is corn or wheat. According to the US Department of Agriculture's estimates of world agricultural supply and demand, in fiscal year 2009–10, 35 percent of the United States' corn output was used to make bioethanol, accounting for about 130,000 square kilometers (50,200 square miles) of arable land, and a year later the share was expected to increase to 39.4 percent, equivalent to 140,600 square kilometers (54,300 square miles).[9] In Brazil, ethanol is produced from sugarcane. Huge areas in the Amazon basin help Brazil meet its energy needs. Brazilian gas stations already sell pure ethanol (known as E100) and 25 percent ethanol-gasoline blends (E25). Overall, biofuels account for 20 percent of the fuel consumed in Brazil. North, South, and Central America combined account for 88 percent of the world's ethanol output (figure 3.2). The EU accounts for only 5 percent of the world's total, with France accounting for 34 percent of the EU production, Germany for 20 percent, and Spain for 12 percent.[10] In Europe, rye and sugar beets are used in addition to corn and wheat.

Biodiesel fuel (often called simply "biodiesel") is a fairly good substitute for fossil diesel fuel, even though 10 percent more volume is needed to generate the same amount of energy. Most modern diesel-engine cars can run on such fuel without any problem. Only a few carmakers have

Bioethanol (2009)   Biodiesel (2008)

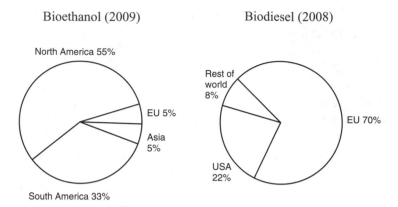

Figure 3.2
Market shares of political regions in worldwide production of bioethanol and biodiesel. Sources: Renewable Fuels Association, *2010 Ethanol Industry Outlook: Climate of Opportunity* (www.ethanolrfa.org); Emerging Markets Online, *Biodiesel 2020*, second edition (http://www.emerging-markets.com).

refrained from approving the use of biodiesel for fear that it could clog fuel lines, injection pumps, and particle filters.

Biodiesel is made from oil-rich seeds, which in Europe means primarily rapeseeds. It can be also made from soybeans, sunflower seeds, and palm oil. In Indonesia, there are palm plantations for producing biodiesel that stretch as far as the eye can see. Attempts to make biodiesel from biogenic waste are also underway.

As figure 3.2 shows, Europe leads the world in the production of biodiesel, with a 70 percent market share. Germany produces 36.4 percent of the biodiesel fuel in the EU, primarily from rapeseeds. With a 26 percent share of the world market in biodiesel production, Germany even comes in ahead of the United States, which has a 22 percent share. Measured in energy units, Europe produced 4.3 times as much biodiesel as bioethanol in 2009.[11] From 1999 to 2009 its biodiesel production increased by 1,748 percent, equivalent to a 34 percent yearly growth rate. For comparison, the United States' bioethanol output in 2008 was 14 times its biodiesel output.[12]

The rise of biofuels is attributable largely to market reactions to rising oil prices in recent years. Nevertheless, government policies (subsidies, blending prescriptions, and infrastructure) have helped to promote bioenergy. In 2005, the US Congress passed the Energy Policy Act, which included subsidies for biotechnology and biofuels amounting to $500 million. Now American energy companies are required to blend ethanol into their fuels. In 2007, for instance, 4.5 billion gallons of ethanol were mandated to be blended in, equivalent to nearly 4.5 percent of US gasoline consumption. This percentage is going to rise over time. In addition, a Farm Bill aimed at the development of bio-refineries was passed in 2002.[13] Brazil consumes a lot of bioethanol, since the government requires energy companies to blend between 20 percent and 25 percent ethanol into their gasoline and subsidizes the production of biodiesel. The Brazilian automobile industry has reacted to these policy measures by developing flex-fuel cars, which are optimized for a blend of gasoline and ethanol.[14]

Most EU countries provide tax reliefs for biofuels and in addition require fuels sold at filling stations to contain certain proportions of biofuels. In Germany, 4.4 percent biodiesel now must be blended into normal diesel fuel, and 3.6 percent bioethanol into gasoline, the percentage rates referring to energy content.[15] An EU directive requires all EU countries to take measures to increase the share of renewables in fuels to at least 10 percent by 2020.[16] All these policy measures of course violate the law of one price discussed in chapter 2. For the reasons explained there they should be replaced with a common price for the emission of carbon from fossil sources by integrating all sectors of the economy into a general emissions trading system.

More than electricity

Bioenergy is by far the most important renewable energy source, contributing 54.5 percent to (final) renewable energy consumption in the OECD countries, more than all other types of renewables together. Arguably it has the largest potential for satisfying the energy needs of the industrialized countries, since it is able to generate nearly perfect substitutes for crude oil, natural gas, and coal which, as shown in figure 2.3, account for 86 percent of the developed countries' final energy demand.

Biomass can be converted to gases that can replace natural gas. These gases are suitable for use in flexible gas and steam power plants and even in appropriately adjusted home furnaces. They can be burned to generate heat or to run engines, even though they have 40 percent less energy per unit of weight than natural gas. They can even be used to produce biodiesel. Methane resulting from the fermentation of biomass is essentially identical to natural gas.

Even more important, biomass can be used to produce liquid fuels that approximate the liquid fuels of fossil origin that power nearly all transportation. Currently, 35 percent of the OECD countries' final energy is from fossil liquid fuels used for transportation on water, on land, and in the air. These could in principle be replaced with bioethanol and biodiesel, which are, as explained above, nearly perfect substitutes, requiring only a bit more volume.

The point here is that all these substances are carbohydrates that are chemically related to hydrocarbons, their fossil kin, and can therefore be used in nearly the same way. After all, fossil energy is also bioenergy that was stored for 300 million years or so in the Earth's crust.

As much as electricity counts as refined and sophisticated energy, in the end liquid and gaseous fuels are more useful for mankind for the simple reason that they can easily be stored and carried around at normal air pressure in simple containers. There is no other energy carrier that can be as comfortably used in mobile engines. For ships, cars, and airplanes, hydrocarbons are ideal energy reservoirs, since they react with oxygen, a substance that is available everywhere and therefore doesn't have to be transported. Indeed, the general availability of oxygen is the big advantage of hydrocarbons. The oxygen needed for combustion is much heavier than the hydrocarbon itself. Burning a kilogram of gasoline requires 2.13 kilograms of oxygen. For a kilogram of diesel fuel, the amount of oxygen needed is even larger: 2.18 kilograms, about the same amount that is necessary for kerosene (2.17 kilograms). Everything that moves on land, on water, or in the air, including animals and humans, needs only to open its mouth or its intake port to add the heavier ingredient to the combustion process; only the lighter ingredient has to be carried around. Only in outer space and under water does this advantage disappear, oxygen having to be carried as well, which is why it is so costly to move things there.

It is true that electric motors are much simpler and efficient than internal-combustion engines. Thus, stationary electric motors outperform stationary internal-combustion engines. However, the electric energy has to be produced from primary energy sources, and transporting it in the form of batteries is cumbersome. Batteries are expensive and complicated, and even the very best ones (the lithium-ion batteries used in laptop computers and in some hybrid cars) have a final energy density per unit of weight that is at best 1,7 percent, perhaps in the future 2 percent of that of diesel fuel. This is partly compensated for by the better efficiency of an electric motor relative to a diesel engine. An electric motor transforms 85–95 percent of the power in its batteries into kinetic energy, whereas even a modern common-rail diesel engine produces at most 45 percent kinetic energy from the diesel fuel it consumes. Taking the two effects together, the kinetic energy obtainable from a kilogram of lithium-ion batteries is about 4 percent of that obtainable from a kilogram of diesel fuel.

And where electricity is preferred, it can easily be made from hydrocarbons by using steam or internal-combustion engines to run generators. In ships, diesel engines are often used to generate electricity that in turn powers the propellers, rather than to move the propellers mechanically. However, hydrocarbons can't be made from electricity at reasonable cost. Filling greenhouse skyscrapers with plants or algae to generate biomass to supplement the biomass made from sunlight is a theoretical possibility, but it is rather absurd from the economic and technical points of view.

Of course, electricity can be used to generate hydrogen by way of electrolysis. Hydrogen is a gas that, upon combustion, can take the oxygen out of the air. It is 62 percent lighter per unit of energy than gasoline. But it isn't a good substitute for fossil fuels, since it can't be stored under ambient pressure and temperature. To be transported, it must be cooled and/or compressed, which requires large and heavy technical devices. Hydrogen compressed to 200 bars (2,900 psi) at a temperature of 25°C (77°F) takes up, relative to its energy content, 15 times as much space as gasoline, and about 15.8 times as much space as diesel fuel or kerosene. Crash-proof tanks that can handle such pressure are so heavy that transporting them more than absorbs hydrogen's higher energy density per unit of weight. An airplane constructed on that basis would be cumbersome, to say the least. In a modern mid-size passenger jet, 5 percent of the volume is used for kerosene, 45 percent for the passenger cabin, and 35 percent for baggage or freight. Replacing the kerosene with 15 times as much hydrogen would take up 75 percent of the space, not counting the massive tank itself. And in a crash, such a tank would be a veritable bomb, even if the hydrogen didn't ignite.

If one cools hydrogen below −250°C (−420°F), so that it becomes a liquid, it requires only four times as much room as gasoline. But then substantial cooling units and insulated containers are necessary, which also rule out many transportation possibilities, particularly in aviation. A significant portion of the energy, on top of that, would be needed to keep the gas cold.

One possibility would be to build zeppelins such as those that toured the world in the 1920s and the 1930s. However, those zeppelins used the hydrogen for lift, not for propulsion. The propulsion was by conventional engines. Someday, perhaps, hybrid flying machines combining features of jet planes and zeppelins will use hydrogen both for lift and propulsion. But because of hydrogen's high flammability, making them safe will pose a great challenge. The era of the zeppelins, after all, ended in 1937 after the biggest zeppelin ever built, the *Hindenburg*, exploded when its hydrogen was ignited by electric sparks.

For all these reasons, electricity from wind, solar, hydro, or nuclear power is a much less suitable substitute for fossil fuels than bioenergy. Bioenergy appears to be the golden path to solving mankind's energy problems.

A dubious eco-balance sheet

Because biofuels give rise to significant greenhouse-gas emissions, they are by no means as environmentally friendly as one might initially think. It is true that the corresponding energy carrier has been previously extracted from the air, so that its combustion has no detrimental effect on the CO_2 content of the atmosphere (provided, of course, that the Earth's stock of biomass remains unchanged). But fertilizing fields with nitrogen leads to a release of nitrous oxide (NO_2), since plants usually can't absorb all the nitrogen provided by fertilizer. As was explained in chapter 1, nitrous oxide is 300 times as potent a greenhouse gas, per unit of weight and over 100 years, as carbon dioxide.

The nitrogen problem could be avoided if fertilizers were continuously metered out and distributed over the field, plant by plant, in just the right quantities that the individual plants can digest. In practice, however, such a continuous and individualized provision is impossible. Inevitably, fertilization takes place in discrete steps covering the entire field, so that there are inevitably long periods of time where the fertilizer is excessive and subject to oxidation. It is impractical to fertilize each plant individually and provide it with only as much nitrogen as it needs for growth[17]; because farmers work with averages, some plants will receive more than they can use. Rain washes away the surplus nitrogen and concentrates it in places where it can either not be used at all or not in the amount available. All these nitrogen surpluses serve as nourishment for bacteria that produce nitrous oxide as waste, which then joins the other greenhouse gases in the atmosphere.

This problem isn't limited to energy crops. It occurs with other crops too. But in the case of energy crops it is particularly strong, because the high prices fetched by such crops encourage farmers to squeeze as much as they can from every square inch of land, even if the marginal gains from adding more fertilizer are small.

A thorough study of the greenhouse-gas effects of bioenergy was conducted in Switzerland in 2007. Figure 3.3 shows one of its findings for the emission of greenhouse gases during biofuel production, relative to the greenhouse gases that are emitted in the production and use (labeled "operations" in the figure) of fossil diesel fuel, natural gas, and gasoline. Identical final-energy content was assumed. The greenhouse-gas emission for gasoline was set at 100 (see the second line in the figure)

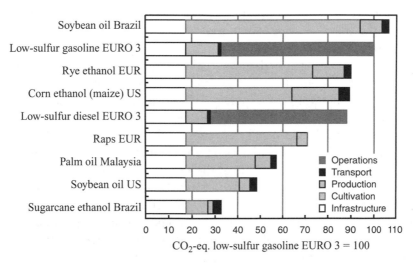

Figure 3.3
Greenhouse-gas emissions resulting from the production and use of different types of fuel, assuming identical provision of final energy. All values are expressed relative to the emissions resulting from the production and use of low-sulfur gasoline satisfying the Euro-3 norm. Source: R. Zah, H. Böni, M. Gauch, R. Hischier, M. Lehmann, and P. Wäger, *Ökobilanz von Energieprodukten: Ökologische Bewertung von Biotreibstoffen*, Swiss Federal Office of Energy, Swiss Federal Office for the Environment, and Swiss Federal Office for Agriculture, Bern, May 22, 2007, figure 79, p. 92.

and was used as the benchmark for comparisons. The light gray bar shows that the largest portion of greenhouse gases in the case of biofuels results from raising the crops. The greenhouse effect is usually so strong that the biofuels lose a large portion of their theoretical comparative advantage over fossil fuels. Particularly poor values are shown by Brazilian biodiesel made from soybean oil, European bioethanol made from rye, and American bioethanol made from corn. In the case of American bioethanol made from corn, by far the most important biofuel sort worldwide, the greenhouse effect is only 10 percent below that of fossil diesel fuel. Biodiesel made from soybeans and rye is negligible by volume. With 70 percent of the emissions from fossil gasoline, European rapeseed lies at or near the middle of the bio fuels. It is noteworthy that Brazilian bioethanol from sugarcane is quite benign for the climate, as very little fertilizer is used in its production. And it is quite benign even though sugarcane is burned before harvesting, in order to make it easier to reach the stems. Thus, with soy beans and sugar cane, Brazil is host to both the most benign and the most damaging biofuels in terms of influencing the greenhouse effect.

The Swiss study resorted for its calculations to established values (also supported by the IPCC) of how much nitrous dioxide the excess nitrogen releases into the atmosphere. Those values, however, are being disputed in the literature. Patzek pointed out in 2006 that much more nitrous oxide is released than is assumed in the calculations.[18] He concluded that the use of bioethanol in the United States is 50 percent more damaging to the climate than the use of petroleum gasoline. The same conclusions were reached by a study conducted by Crutzen and colleagues.[19] Crutzen et al. also find a greenhouse effect 50 percent above that of fossil fuels. For European biodiesel produced from rapeseed, they arrive at a greenhouse effect 70 percent higher than that of normal diesel fuel. On the other hand, Liska et al. argue that the pessimistic outcomes of the above studies largely resulted from the fact that they referred to traditional methods of ethanol production.[20] New methods available today have a much higher energy efficiency and hence will indeed entail lower CO_2 emissions relative to fossil fuels. The new bio-refinery plants, for example, use thermo-compressors for condensing steam, thermal oxidizers for the combustion of organic compounds, a system for recovering heat that would have been wasted, and raw-starch hydrolysis that reduces the heat requirement during fermentation. Liska et al. maintain that these components, taken together, would reduce the climate-damaging gases by two-thirds relative to fossil fuels.

The BtL hope

It is hoped that second-generation biofuels will be a real breakthrough. First-generation biofuel production used only the oily, starchy, or sugary seeds of the feedstock plants and could at best burn the other parts to produce heat. But in principle it is conceivable to use the entire plant—including the woody portions known as lignocellulose frame—for the production of biofuel, significantly improving the energy output at the same volume of nitrous oxide released. Burned in a kiln, the plants' woody parts contribute with a significant portion, if not most, of the combustion energy. Second-generation fuels liquefy the energy contained in the woody parts of the plants and make it available for use in engines. The technical term here is *biomass-to-liquid* (BtL). There are two ways to accomplish this liquefaction.

One way is to use microbiological fermentation to produce bioethanol from the whole plant. Pilot plants are in operation in Ottawa, Canada, and in Jennings, Louisiana. The Ottawa plant, run by the Iogen Corporation, can process up to 30 tons of straw per day and should produce about 2 million liters (that is, half a million gallons) of ethanol per year. The plant in Jennings, run by the Verenium Corporation, is to have an output exceeding 5 million liters (1.3 million gallons) per year, and Verenium plans to erect other, larger plants. The Ottawa and Jennings plants break down the plant matter using acids, and then apply special enzymes to transform it into various kinds of sugars, which are then fermented into ethanol using yeast. Not all the sugars produced in this way lend themselves to being transformed into ethanol by simple yeast. This has prompted researchers to try genetic engineering to develop other kinds of yeast that can ferment a larger spectrum of sugars into ethanol.

The other way to accomplish liquefaction is to produce synthesis gas from plants and then refine it into a diesel-like fuel. The method of producing liquid fuel from synthesis gas was devised in 1925 in Germany by Franz Fischer and Hans Tropsch and is thus known as Fischer-Tropsch synthesis. In the 1930s, large quantities of liquid fuel were produced in Germany's Ruhr Basin. Later, the Fischer-Tropsch method was used extensively in the former German Democratic Republic. It is still used today in South Africa in the production of fuels from brown and hard coal. In Freiberg, Germany, the method is now used for the production of fuel from biomass in the Carbo-V process.[21] First the biomass is converted into synthesis gas, then the synthesis gas is turned into diesel fuel and naphtha. Naphtha is a by-product used in the chemical industry to produce polyethylene, polypropylene, and other materials.

Synthesis gas doesn't have to be converted to liquid fuel to be useful for the generation of energy. It can also be used to run engines. Before World War II, trucks running on thermally obtained gas were not uncommon in Germany. A box attached to the external part of the vehicle was filled with wood, which was burned at low oxygen levels and gasified, producing the synthesis gas. This gas was routed to the truck's engine and used to power it.

BtL fuels could represent a breakthrough for bioenergy, as their production doesn't require energy crops. All kinds of feedstock, from slaughterhouse waste to straw, can be used.[22] Utilizing the woody parts of plants results in a much higher energy output than using mere energy crops. This would make it possible to crowd out a higher volume of fossil fuels.

The BtL technology is still in its infancy, however, and what its real market potential will be isn't yet clear. There are concerns that the biological equilibrium of land cultivation would be destroyed.[23] Nevertheless, this technology is an elegant way to circumvent the ethical issues surrounding the table-or-tank debate and to develop a bio-energy strategy that focuses directly on the utilization of biogenic waste.

Slash and burn

Alongside the bright spark of BtL there are practices that cast a murkier light on bioenergy's ecological credentials. These include the depletion of the world's current biomass stock. As was pointed out above, resorting to bioenergy is climate neutral only when the carbon bound in biomass isn't burned any faster than would have occurred through natural oxidation. When jungles are cleared to make room for fuel crops, this condition is blatantly violated, as the cleared fields will carry a permanently lower amount of biomass afterward. The original stock of tall, older trees is usually replaced by smaller plants that are harvested yearly and thus bind only a minimum amount of biomass to the fields in question.

If the wood obtained from clearing the jungle is used as firewood, replacing fossil fuels, the forest clearance would be, in itself, CO_2 neutral. But that is rarely done. Instead, the land is often cleared by slashing and burning the forests. Indonesia and Malaysia, for example, clear large swathes of rainforest to make room for oil-palm plantations.

In Brazil the situation was similar. Large parts of the Amazon rainforest were burned to clear land for the production of sugarcane. The situation has changed, the Brazilian government having imposed strict zoning regulation that forbids the direct conversion of rainforest into sugarcane production. Thus, about one-third of the recent expansion of sugarcane fields occurred on former cropland instead, and the remainder was gained by converting pastures.[24] It is debatable, however, whether the use of former cropland really made a difference, because the land for growing food crops had to be replenished from other sources. To a large extent, the ongoing destruction of the rainforest for the purpose of gaining cropland may have been a mere substitution effect brought about by market forces: the production of ethanol on former cropland increased the prices of crops, making it increasingly attractive to clear more rainforest.

Consider one hectare (2.5 acres) of a Brazilian sugarcane plantation. The sugarcane grown there is processed into bioethanol that is used to replace fossil fuels in proportion to its energy content, thereby reducing global CO_2 emissions year after year. Suppose this hectare replaces a hectare of cropland, which elsewhere replaces a hectare of rainforest cleared by burning. Burning the forest releases the carbon bound in the vegetation and the soil. The amount of carbon generated by burning is a "carbon debt" that has to be redeemed by replacing fossil fuel with biofuel produced on the cleared land in the future. How long must one wait until the carbon debt is paid back? In the case of the Brazilian rainforest the debt amounts to 737 metric tons of CO_2 per hectare.[25] The ethanol produced from the sugarcane replaces fossil gasoline, reducing the carbon debt by 9.8 metric tons per year. Thus, it would take 75 years to pay back the resulting carbon debt.

It looks even worse in the case of Indonesian or Malaysian rainforest being burned to make room for oil palms to produce biodiesel fuel, 86 years being necessary to repay the carbon debt. If the rainforest burned was growing on peaty soils that must be dried in order to grow crops on them, so much additional carbon dioxide and methane will be released that it will take 423 years to offset the effect of clearing in this case. With biodiesel from Brazilian soybeans, the figure is 319 years. A more

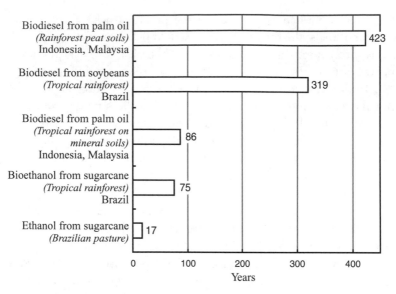

Figure 3.4
Ecological amortization period after fire clearing. Numbers represent years it would take to offset the carbon dioxide released by burning a certain area of forest, through the lower emissions resulting from using fuels derived from energy crops raised on the land in question, under the assumption that the fossil fuels displaced by these bio-alternatives will remain underground. Sources: J. Fargione, J. Hill, D. Tilman, S. Polasky, and P. Hawthorne, "Land clearing and the biofuel carbon debt," *Science* 319, 2008: 1235–1238; calculations by H.-W. Sinn. The calculations by Fargione et al. were complemented for Brazilian sugarcane grown on rainforest-cleared land by consistently applying the data contained in the same article.

favorable balance can be obtained with sugarcane grown on former pastures like the Brazilian Cerrado, as pastures don't carry much biomass either on them or in their soil. In this case, the amortization period is 17 years. Figure 3.4 depicts these findings.

Taken as a whole, these findings show that growing energy crops on soils that first have to be cleared by burning brings no advantages for the environment over many decades. Such strategies should be ditched forthwith if we are serious about stopping climate change.

One hectare for me!

Biofuels don't have to be obtained from areas cleared of jungle. They can also be obtained from crops grown on existing arable land, or on marginal land, or, as mentioned above, from slurry, slaughterhouse waste, straw, and the like. In practice, however, so far existing arable land plays the dominant role. Waste accounts for only 11 percent of bioenergy in Europe and 12 percent in the OECD countries. As a rule, waste is burned or fermented to produce methane, which is then used to generate electricity and for heating. How much BtL will change this remains to be seen.

Advocates of biofuel place much hope on marginal land. In fact, many geological studies point to apparently fallow lands and unused pastures in developing countries that could be used for the production of biofuels. Perennial crops that would sequester carbon in their roots and in addition produce oil seeds, such as jatropha, offer promising options. However, such lands are typically not easily usable, because of a lack of sufficient moisture, accessibility issues, and the risk of destroying wildlife.[26] Currently, almost by definition, marginal land plays no major role, on account of the intensive cultivation required by energy crops. Nearly all the land used for cultivating energy crops is taken away from food production.

Of course, things may change, as increasing energy needs will induce mankind to seek more and more land in remote places. However, this will take place only if rising food prices lift land that currently is marginal and unprofitable above the profitability threshold, that is, only when food land is converted into fuel land. The possibility of expanding croplands for the production of biofuels mitigates the fundamental conflict between using land for food or energy production, but it doesn't resolve it, since the expansion of cropland will only take place if and because food prices increase.

To get a feeling for the magnitude of the conflict, let us do some calculations on the basis of the arable land currently in use. If it turns out that only tiny amounts of that land will be needed to produce substantial amounts of bioenergy, there is hope that a moderate price increase will be sufficient such that not very much marginal land will be brought in. If, on the other hand, the land use is substantial, a more deleterious effect on food prices is to be feared, even though some of the land needed will come from the fallow land reservoirs available in Earth.

As was shown in figure 2.1, fuels are used primarily for transportation. Let us therefore concentrate on that and investigate, first, how much land would be needed to power a car. At first sight, the relevant calculations show a manageable area. One hectare (2.5 acres) planted with rape can produce 3.4 metric tons of rapeseed, which can be processed into

1,550 liters (410 gallons) of biodiesel, equivalent to 1,410 liters (370 gallons) of petroleum diesel fuel. If we assume fuel consumption of 17 kilometers per liter (about 40 miles per gallon)—good performance for a modern compact diesel car—this would make it possible to drive approximately 15,000 miles (23,500 kilometers), more than the average number of miles a car is driven per year. "One hectare for me!" could well be the battle cry to achieve the bio-turning point.

A similar picture results if the crop is corn instead of rape and the product is bioethanol instead of biodiesel. The output is about 2,560 liters (680 gallons) which, owing to ethanol's lower energy content, corresponds to 1,660 liters (440 gallons) of gasoline or 1,500 liters (400 gallons) of diesel fuel. With an average fuel consumption of 13 kilometers per liter (31 miles per gallon)—realistic mileage in view of the lower energy content of gasoline and the state of engine development—this amount of biofuel would yield about 22,100 kilometers (about 13,700 miles) a year. Table 3.1 summarizes the data.

No matter how reasonable this calculation appears, it becomes quickly sobering when we consider how much land would be necessary to meet the fuel needs of a modern country. The International Energy Agency (IEA) did the calculations for the entire world and came up with a very simple formula: For an x percent blend of biofuel in total liquid fuel consumed, approximately x percent of the world's arable land would be needed. Thus, with a biofuel share of 10 percent per unit of energy the land needed would be just 10 percent of the world's total arable land. As was explained above, this calculation isn't an unconditional forecast; it is a fictitious calculation abstracting from the cultivation of additional marginal fallow land, which could be induced by increasing food and

Table 3.1
How far can a car run on one hectare? Assumptions: fuel consumption of 17 kilometers per liter of diesel fuel (40 miles per gallon) and 13 kilometers per liter of gasoline (31 miles per gallon). Source: Fachagentur Nachwachsende Rohstoffe, *Biokraftstoffe Basisdaten Deutschland*, 2008.

Fuel	Yield (liters/hectare/year)	Range (kilometers/hectare/year)
Biodiesel	1,410 (diesel equivalent)	23,500
Bioethanol	1,660 (gasoline equivalent)	22,100

Table 3.2
Share of bioenergy in fuels and share of required arable land (IEA approach).
Sources: International Energy Agency, *World Energy Outlook 2006*; Eurostat, *Energy—Monthly Statistics*, August 2007; Renewable Fuels Association, *2010 Ethanol Industry Outlook: Climate of Opportunity* (www.ethanolrfa.org).

Biofuels	World	US	EU 15
10%	10%	9%	31%
20%	20%	19%	62%
100%	100%	94%	308%

land prices. In economic terms, it is a calculation of excess demand for land at existing prices induced by biofuel production. Any model trying to predict endogenous, price-induced reactions of other economic variables needs the excess demand for land induced by biofuel production at given prices.

Table 3.2, which extends the IEA approach to the EU 15 countries and the United States, is based on these countries' specific fuel-consumption patterns, assuming a similar mix of soil suitability as in the world as a whole. A 10 percent share of biofuels in the EU 15 countries' liquid fuel consumption, which the EU has mandated for the year 2020, would require about 21.5 million hectares (54 million acres). This represents 31 percent of all the arable land available there.[27] In the US, on the other hand, only 9.4 percent of the arable land would be needed for a 10 percent biofuel share in overall fuel consumption.[28]

At first glance the number cited for the European Union doesn't agree with the official view held by the EU Commission that a 10 percent biofuel blend would require just 15 percent of the arable land of the (total) European Union.[29] However, as was explained by researchers who redid the EU calculations, the EU figure rests on the assumption of a substantial production of bioethanol from waste and, in particular, on the assumption that half of the biofuel used in the EU is imported from other regions of the world.[30] Removing these two assumptions would imply devoting more than 30 percent of the arable land to such energy crops, which is consistent with table 3.2.

The reason for the relatively high land requirement in the European Union is the higher population density, which leaves less land free for cultivation. The US figure comes relatively close to the world average, as the pattern of its land use is quite similar to the world average.

The dream of being able to do without fossil fuels in the future—that is, having the transport sector, which accounts for about a fifth of primary and nearly a third of final energy consumption in the world, run entirely on biofuels—is obviously just that: a dream. At existing prices this would require all of the world's arable land to be devoted solely to biofuel production. The EU would require about three times as much arable land as it has, and in the United States it would require 94 percent of all arable land, leaving just 6 percent to feed the population.

One day, perhaps, the BtL processes will help to bring these figures down, but BtL is still in the experimental phase. If, as promised, BtL extracts yields twice or three-times higher than today, the required arable land area will shrink by one-half to two-thirds. This makes a better balance for the world and the United States, but the European balance would still be horrifying. The European cropland would barely suffice to produce the fuels required for transport. Food crops, naturally, would fall by the wayside in this case.

It is debatable whether the nitrous dioxide problem resulting from fertilization implies that biofuels are more damaging to the climate than the fossil sort. However, it is undeniable that the rainforests being fire-razed in developing countries to make room for energy crops bring about a carbon debt whose redemption through the usage of biofuels will take decades, even under the most favorable circumstances. And the amount of arable land required to produce biofuels is so enormous that even if BtL fulfills its promise, little would be left for food crops, except perhaps at horrendous prices. The only conclusion is that the biofuel idea, unless strictly limited to the use of biogenic waste, is an error in the history of humanity.

As was explained above, all renewable energies are essentially attempts to collect and concentrate the flow of energy coming from the sun, which is gigantic but thinly dispersed over the Earth's surface, and they must show how they can accomplish this with the lowest possible opportunity costs. Bioenergy seemed to be a particularly promising candidate, since it doesn't require technical installations covering vast areas and can instead resort to existing agricultural production methods, which are designed to collect the sunlight over extensive areas of land in a way that is quite environmentally acceptable, at least one mankind is accustomed to. However, what at first glance seemed to be an advantage turns out on closer scrutiny to be the biggest disadvantage. The gigantic requirement of arable land creates the problem for nutrition. We would starve if we were to insist on running our vehicles only on bioenergy.

Farmers to OPEC!

Despite this bleak forecast, production of biofuels is rising dramatically, as figure 3.5 shows. From 2000 to 2009, global bioethanol production increased at an average annual rate of 17.9 percent, jumping from nearly 9 million to 38 million tons of oil equivalent. Production of biodiesel went from nearly nothing to about 10 million tons of oil equivalent.

There were three forces behind the biofuel surge in the first decade of this century. The first was political. The "green" movement gained momentum in many European countries and in the United States. The idea of producing energy from biomass without burdening the carbon cycle with more fossil fuel was tempting enough to gain strong political support. The Kyoto Protocol, which became effective in 2005, encouraged the use of biofuels to meet the emission targets. In addition, after the attack on the World Trade Center on September 11, 2001 and the invasion of Iraq in 2003, there was a growing awareness in the United States of the country's energy dependence and the supply risks involved in importing oil from Arab countries. This contributed to political decisions that paved the way toward bioenergy. Policy measures—blending requirements, tax reliefs, subsidies, infrastructure investment to facilitate the conversion of crops to fuel and the like—were taken precisely for these reasons in many countries.

The second force was energy companies. Because of the sharp rise in oil prices during recent years, they became increasingly interested in alternatives to fossil fuels. They built huge production facilities for the refinement of biofuels in the United States and Europe, thereby increasing their demand for biofuels dramatically.

The third force was, of course, farmers, who smelled a business opportunity. New customers are always welcome. Profits go up in any case, either because more quantities are produced or, if that isn't possible, because prices go up. That new customers would crowd out old customers was a problem for the old customers, but not for the farmers. It is true that redirecting their attention from the food market to the oil industry wasn't quite what their traditional values would have suggested. But when enough money flows, adjustments come quickly. Helmut Born, secretary-general of Deutscher Bauernverband (German Farmers'

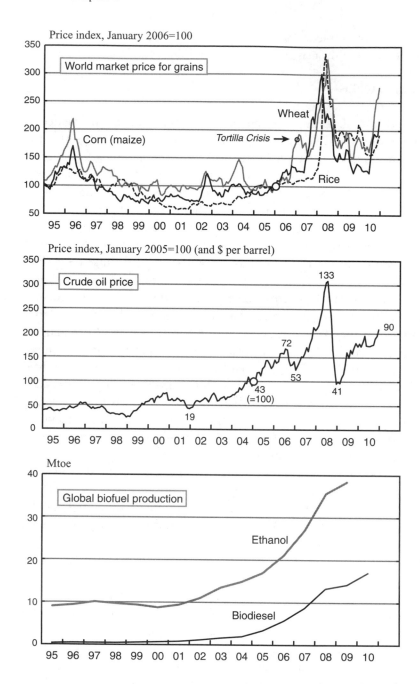

Figure 3.5
Food prices, oil prices, and biofuel production—causality or coincidence? Topmost panel shows monthly world prices for wheat, rice, and corn as an index, for which January 2006 was taken as the baseline in order to depict more clearly the increase in prices until the Tortilla Crisis of 2007. Middle panel shows monthly price of crude oil per barrel, again as an index curve. Here the base is January 2005. The numbers attached to the curve are dollar prices per barrel. Bottom panel shows evolution of ethanol and biodiesel production in oil-equivalent energy units. Mtoe stands for million tons of oil-equivalent. (One kilogram oil-equivalent corresponds to 11.63 kilowatt-hours.) Sources: Hamburgisches WeltWirtschaftsInstitut, *Rohstoffpreisindex* (http://www.hwwi-rohindex.org); Earth Policy Institute, *World Biodiesel Production 1991–2010* (http://www.earth-policy.org); J. von Braun, *Promises and Challenges When Food Makes Fuel*, International Food Policy Research Institute, 2007; BP, *Statistical Review of World Energy—Renewables*, 2010.

◄───────────────────────────────────

Association), argued that farmers should get accustomed to the idea that from now on food prices will increase in tandem with the price of oil.[31]

Born may be right. In the last decade, at least until the worldwide economic crisis of 2008, oil and food prices were not only increasing; they were exploding. As figure 3.5 shows, the prices of corn, wheat, and rice soared together with the price of oil in 2008. The visual parallels between the evolution of oil and food prices is striking indeed. Farmers had every reason to celebrate on a grand scale. Perhaps they should even pay a membership fee to the Organization of the Petroleum Exporting Countries.

Of course, coincidence doesn't mean causality, not to mention monocausality. The world's growing population, the increasing demand for higher-value food (meat instead of grain), income improvements around the world, and, not least, bad weather spells all contributed to push grain prices up. Higher oil prices also played a role insofar as they increased the cost of growing grain. Still, the fact that the increasing demand for biofuel withdrew land from food production and made food scarcer, thus increasing food prices, was a major factor.

Bioethanol production started in the mid 1970s as a reaction to the 1974 oil crisis. After picking up pace during the second oil crisis (1980–1982), it advanced little in years 1985–2000, when the price of oil was fairly constant. Then, around 2001, oil prices began to climb steadily, exceeding $130 per barrel in 2008. Biofuel production increased simultaneously. Between 2000 and 2008, oil prices quadrupled, and so did bioethanol production. Production of biodiesel increased more than twelvefold. In the United States, more and more farmland was turned from wheat to corn production, and ever more corn was used to make bioethanol. Europe, and in particular Germany, acquired a distinct yellow hue, as rape was grown everywhere to produce biodiesel. Until 1975, nearly all of the EU's farmland was used for food production. By 2009, 17 percent of the farmland in Germany and 7 percent of that in the EU's 27 member states was used for biofuel production.[32] This was due in part to the policy measures described above in the section on "green" gasoline, and in part to the increase in the price of oil.

Much more important than what happened in Europe were developments in the United States, which accounts for 40 percent of the world's corn output and 70 percent of the world's corn exports.[33] Nearly all of the increase in the world's corn production from 2004 until 2007 ended up as input for the United States' ballooning bioethanol production.[34] Such crops accounted for 8 percent of the cultivated area of the US in 2007, and 30 percent of the overall US corn output was then used for producing bioethanol.[35] As was mentioned above, the US Department of Agriculture even estimated a share of 35 percent for the fiscal year 2009 and a share of nearly 40 percent for the fiscal year 2010.

One reason for the boom in bioethanol production was the increase in the price of oil, which made the production of bioethanol more profitable. Another reason was a public policy that helped to couple the food and energy markets. As early as 1978, the Energy Tax Act had exempted gasoline with 10 percent ethanol content from the gasoline tax.[36] The Energy Policy Act of 2005 created tax incentives for bioethanol facilities and for the production of 15 percent bioethanol-gasoline blends.[37]

The Energy Independence and Security Act of 2007 allowed for another increase of corn-starch ethanol production to 15 billion gallons until 2015—an increase of 131 percent relative to the 2007 production volume of about 6.5 billion gallons. However, it also stipulated that further production increases beyond 2015 are to be generated from "advanced biofuels" of the BtL type, which must not be produced from corn starch.[38] The Food, Conservation, and Energy Act of 2008, furthermore, specified subsidies that are tailored to the production of bioethanol and biodiesel from biomass other than corn starch.[39] It remains to be seen whether this policy change will be able to stop the momentum toward ever-increasing bioethanol production from corn after 2015.

Using corn to make bioethanol turned the world's food market on its head, affecting other markets through substitution effects on both the demand side and the supply side. For one thing, farmers around the world switched from growing wheat to growing corn, as the business case for the latter was more compelling. The growing scarcity of wheat then pushed the price of wheat upward. And when corn and wheat consumers tried to escape the price increases by switching to rice, the price of rice went up.

According to a study by the International Food Policy Research Institute (IFPRI), biofuel production accounted for 30 percent of the rise in the average price of wheat, corn, and rice between 2000 and 2007. In the case of corn, it accounted for 39 percent of the price increase.[40] Corn prices were one-third higher in 2007 than they would have been without biofuel production. Nevertheless, the IFPRI study argued that 61 percent of the price hike that occurred over that period was attributable to structural factors such as population growth and changes in nutrition habits.

Even more alarming findings were published in a World Bank paper authored by Donald Mitchell, one of the World Bank's senior agricultural economists. Examining the period from January 2002 to February 2008, Mitchell found that only one fourth of the corn price increase could be attributed to dollar depreciation, higher production costs, and trend phenomena. Most or all of the remaining three-fourths resulted from the conversion of arable land to biofuel production.[41] Mitchell's calculations took account of the effect of speculation-driven scarcity and the export bans introduced by certain producing countries, which in his opinion were themselves caused by the price increases. Such bans were applied by Argentina, India, Kazakhstan, Pakistan, Ukraine, Russia, and Vietnam with the purpose of protecting their populations from food scarcity.

The findings of Mitchell and IFPRI have been intensely discussed both in public policy debates[42] and in the academic literature. Some of the academics, including Piesse and Thirtle[43] and Headey and Fan,[44] have confirmed Mitchell. But more skeptical voices have raised doubts about the causal link between the price of oil and that of food. In particular, Gilbert's assertion that the correlation between the prices resulted from a common demand effect explaining both prices rather than a causality from oil prices to food prices has drawn much attention recently.[45]

The differences in opinion, however, are less fundamental than they may seem at first glance. No one disputes the strong correlation between food prices and energy prices, and of course no one has argued that the former affected the latter. The point is simply that it is hard to distinguish empirically between a clear causality from oil prices to food prices and a co-determination of both prices by a general increase in energy demand. In fact, if corn-starch ethanol and gasoline are nearly perfect substitutes and if there is therefore a particularly firm link between the food and energy markets on the demand side, the causality question loses its meaning.

If anything, the econometric findings make the influence of political decisions, whether justified with environmental protection arguments or with national supply security arguments, fade into the background. Perhaps these decisions were less important than they appeared at first sight, or, more probably, they were themselves endogenously induced by rising oil prices and the resulting pressure from farmers' lobbies and energy lobbies that scented good business.

A cursory glance at figure 3.5 confirms the close correlation between the prices. From January 2005 to the temporary peak in the summer of 2006, the price of oil increased by 67 percent. With a delay of about half a year, this increase led to a temporary peak in the price of corn in January 2007, which in turn triggered the Tortilla Crisis. Corn prices shot up by 83 percent between January 2006 and January 2007. Wheat prices followed exactly one harvesting cycle later, as farmers had switched from growing wheat to growing corn, making wheat scarcer and therefore more expensive. As consumers switched from expensive wheat to rice, the price of the latter inched up as well. This happened relatively quickly, a few months later. The large increase in the supply of corn then retarded the corn price hike, but the higher prices for wheat sent a signal to farmers that they wouldn't have to switch to corn to increase their revenues, prompting many to return to wheat production. This, of course, had the effect of pushing corn prices up a second time, albeit with a delay. The process repeated itself until the prices reached their peaks in early 2008. The sufferers were the consumers, in particular those who couldn't afford food and thus were threatened by hunger.

The average monthly oil price reached a peak of $133 per barrel in July 2008, then fell sharply as the financial crisis turned into a real economic crisis, pushing the world's economy into recession in the years 2008 and 2009. It bottomed out at approximately $41 per barrel in December 2008. The prices of wheat, corn, and rice plummeted simultaneously. However, the period of low prices was very limited. The gigantic state rescue measures taken by world's governments —a $7 trillion bank rescue program and a $1.650 trillion Keynesian recovery program—allowed the world's economy to recover quickly.[46] By December 2010, the price of oil had risen to $90 dollars a barrel. With it, the prices of corn, wheat, and rice were also on the rise, with corn prices again taking the lead.

The Tortilla Crisis

Between 2005 and 2008, the rising price of oil broke the barrier between the oil market and the food market. Two hitherto separate markets became entwined, and carbon leaked from the food to the energy market, making food more scarce and expensive. Human nutrition is based on carbohydrates, all of which consist of carbon compounds that are quite similar to those of the fossil fuels that power our engines, power stations, and heating systems. In this light, the coupling of the two markets responds to a reasonable market logic. But that is no consolation for those affected. The world's poor see the rich putting in their tanks what they need on their tables. If new barriers don't stop this development, conflicts all over the world are inevitable.

A foretaste of what may come was provided by Mexico City's Tortilla Crisis of 2007.[47] Tortillas are a staple in Mexico, and the country imports much of the corn it needs to make them from the United States. With the diversion of corn to bioethanol production, the price of corn shot up, and with it the price of Mexican tortillas. The near doubling of corn prices between the winter of 2005–06 and the following winter caused tortilla prices to rise by 35 percent. That gave rise to huge street protests.

But the true crisis came a year later when the prices of wheat and rice reacted to the increase in the price of corn. Many more people were affected by this than by the increasing corn prices alone. In the first months of 2008, hunger ravaged nearly all developing countries. Haiti, Indonesia, Egypt, and Senegal were particularly affected. The problem spread all over the world, as all countries were linked through the common world market for foodstuffs. Hunger protests, many of them

Table 3.3
Major food protests in 2007 and 2008 as reported in the German press. Sources: For Haiti: AG Friedensforschung der Uni Kassel, *Aufruhr im Land der Berge,* 2008 (http://www.uni-kassel.de/). For Egypt: "Egyptian workers riot over rising prices," *therawstory*, April 6, 2008 (http://www.rawstory.com). For Ivory Coast, Honduras, Mauritania, Cameroon, Senegal, Burkina Faso, Yemen, Mozambique, India and Indonesia: "Droht uns eine globale Katastrophe?—Hungerproteste in aller Welt," *Die Zeit*, June 3, 2008 (http://www.zeit.de). For Peru: "Pfeile gegen die Regierung," *Süddeutsche Zeitung*, July 10, 2008 (http://www.sueddeutsche.de). For Bangladesh: "Sandkastenliberale üben Schadensbegrenzung," *WOZ—Die Wochenzeitung*, April 17, 2008 (http://www.woz.ch).

Country	Date	Number of protesters	Casualties
Haiti	April 7, 2008	Several thousand	At least 4 dead, more than 30 injured
Egypt	April 6, 2008	25,000	80 injured
Ivory Coast	Early April 2008	1,500	At least 1 dead
Mexico	January 2007	75,000	—
Peru	July 9, 2008	6,000	N.A.
Honduras	April 17, 2008	Tens of thousands	N.A.
Mauritania	November 2007	1,000	6 dead
Senegal	Late April 2008	Several thousand	N.A.
Burkina Faso	April 2008	N.A.	N.A.
Cameroon	March 2008	N.A.	At least 100 dead
Yemen	Late March 2008	N.A.	N.A.
Mozambique	February 2008	N.A.	6 dead
Bangladesh	April 12, 2008	20,000	N.A.
India	April 21, 2008	Several thousand	N.A.
Indonesia	January 2008	10,000	N.A.

violent, occurred in 37 countries. Table 3.3 gives an overview of the events from January 2007 to April 2008, when the protests culminated.

As oil prices rise, mankind will have to get used to such events in less developed countries, and political leaders in the Western world will have to avoid making policy decisions in the sphere of energy and agriculture that will one day destabilize the planet. If worldwide hunger riots lead to civil wars in the emerging countries, a wave of revolution may sweep around the world.

The Jasmine Revolution that broke out in Tunisia in December of 2010 and spread to Egypt, Libya, Yemen, Bahrain, and Syria in the following months was only superficially a political revolution to achieve freedom and democracy. A deeper motive was the desperate economic situation of the population in the Arab countries, particularly among the young who had no jobs and no hopes for the future. The quickly spreading hunger during the winter of 2010–11, when domestic food supplies became scarcer and imported staples such as rice and wheat became increasingly more expensive, was among the forces that drove people to the street and triggered violent actions. Figure 3.5 shows how rapidly food prices were increasing toward the end of 2010, following the new oil price hike after the world economy recovered from the 2009 recession. In the winter of 2010–11, the food price index published by the UN's Food and Agriculture Organization (a composite indicator for oils and fats, sugar, cereals, meat, and dairy products) climbed to about the same level as in the early winter of 2007–08; it reached its highest historical value, significantly above the peak that triggered the Tortilla Crisis, in February of 2011.

It is true that globalization has brought many benefits to the world's poorer countries. It has reduced the disparities between national labor markets, and it has induced wage convergence that has dramatically reduced poverty around the world. From 1981 to 2005, the fraction of the world's population living below the \$1.25-per-day poverty line set by the World Bank declined from 52 percent to 25 percent.[48] However, as the hunger crisis showed, globalization has also made the poor vulnerable to increases in food prices caused by the rechanneling of crops from their tables to fuel tanks. They are rightly unwilling to accept this. The Tortilla Crisis may have been a harbinger of widespread conflict and unrest to come.

The Ratchet Effect

The coupling of oil and food prices occurred in the past few years with astonishing force and abruptness. Apart from the political influence, which probably was endogenous anyway, the abruptness can be explained by the fact that crude oil and foodstuffs aren't substitutable in both directions: foodstuffs can be turned to fuel, but fuel can't be converted to foodstuffs, at least not in a direct, chemical sense. We can't use the oil we buy at a filling station to fry food or to spice up a salad, and even with sophisticated bio-chemical methods we can't turn it into animal feed. However, when the price of diesel fuel peaked in 2007 and 2008, some people bought vegetable oil in grocery stores and ran their old diesel cars on it. Modern diesel cars have trouble running on vegetable oil, but chemistry can easily circumvent them and convert the vegetable oil to biodiesel. Most gasoline engines can tolerate a certain blend of bioethanol, and Brazilian flex-fuel engines can even run on pure ethanol.

This one-sided substitutability implied that the energy equivalent prices in the two markets couldn't converge when the price of oil still hovered below the price of food, measured in energy equivalents. It was tempting to convert crude oil to food, but all chemists who tried to do it failed. The situation changed dramatically when all of a sudden the price of oil caught up with the price of food and even hinted that it would go above it. In that situation, the two markets became coupled, as farmers and energy companies were able to make profits by using energy crops as feedstock for the production of gasoline and diesel fuel. Oil prices were linked to food prices in a way akin to the link between a bicycle's pedals and its rear wheel. If you let the bike coast, the pedals are uncoupled from the wheel. But if you pedal faster, the gears engage and the force applied to the pedals is transferred to the wheel. It is a sort of ratchet effect.

In the aftermath of the worldwide financial crisis, with sharply declining oil prices, the bike was coasting once again, but this was only a temporary phenomenon, disappearing when the world's economy recovered and oil prices perked up again. In all likelihood, the coasting periods for food prices will become increasingly rare in the years and decades to come, if they don't disappear altogether, as oil prices, because of the finite nature of that resource, will trend ever upward, exerting increasing upward pressure on food prices.

In real life, of course, this coupling isn't as rigid as that between a bike's pedals and its wheel. But one need not be an economist to see that the farmers and energy companies will have an incentive to redirect land from the production of food to the production of biofuels if the price of oil tends to exceed the price of food. And it is also easy to understand why this implies that the price of oil will pull the price of food along, while the reverse can't be the case because fossil oil tastes bad on a salad. In view of this, we should indeed be afraid that our farmers will be tempted to hold hands with OPEC.

Strong opposition to this trend can be expected from developing countries, where hunger will increase and with it the potential for political conflict. The dimensions of the problems this will bring can only be guessed at, but it is clear that with the Tortilla Crisis humanity entered a new chapter in its history.

A tale of carbon and man

The coupling of the food and energy markets isn't an entirely new phenomenon in history. In fact, those markets always were coupled until the middle of the eighteenth century, when the Industrial Revolution began. Apart from water and wind, kinetic energy that didn't come from human muscles was always provided by animals that consumed bioenergy. Transportation on land, in particular, was powered entirely by animal muscle. People used horses, camels, oxen, and other suitable animals to move themselves, their trading goods, and their implements, and large tracts of arable land were needed to feed those animals. Today, transportation (including the movement of rolling agricultural machinery) absorbs about 30 percent of the world's final energy use. Before industrialization, the percentage of land that was used to grow fodder for draft and pack animals may have been close to 30 percent, or even higher.

In Bavaria in 1873, according to a detailed historical account of land use, one-third of agricultural output, net of seed crops, was used to feed the draft and pack animals used for agricultural production itself, and it is estimated that another tenth was used for to feed farm animals whose services were sold to other sectors of the economy.[49] With the horses owned and employed outside agriculture for transportation services included, the percentage of the land that was used to feed draft and pack animals may well have been 50 percent or more—that at a time when railroads already were in operation and steamships were transporting heavy freight over rivers and canals. Over time, the share of land devoted to the generation of energy for transportation decreased significantly, but even in 1920, after the spread of railroads and electric motors but before the large-scale use of tractors, one-fourth of the arable land in the United States seems to have been ultimately used for transportation purposes.[50]

The days when land was used for both food and fodder were grim, from today's perspective, because there wasn't enough biocarbon for both purposes. Mankind was caught in the Malthusian population trap.[51]

Put simply, the Malthusian population law says that when the population grows more rapidly than the food supply, nutrition per person diminishes and a larger proportion of the offspring die. If the population grows more slowly than the food supply, or declines, the per-capita food supply increases and more of the offspring survive. On average, enough off-spring survive to maintain the balance between the population and the available means of livelihood. There can be only as many people as the available food, air, and room can accommodate. The Scotsman Adam Smith described it in a famous dictum: "Every species of animals naturally multiplies in proportion to the means of their subsistence, and no species can ever multiply beyond it."[52]

During the Malthusian period, any increase in population always absorbed improvements in the standard of living that could have resulted from capital accumulation, technological progress, and, in particular, the continuous improvement in the division of labor. The relentless competition between economic progress and population growth brought about recurring bouts of starvation in Europe until the nineteenth century. The period from about 1750 to 1850 is referred to as the period of pauperism, because strong population growth had impoverished the populace.

A particular difficulty was that the division of labor and the increased trade it entailed required more transportation services. Though sailing was possible with wind power, transportation on land required more horses and more arable land to feed them, which tightened the Malthusian constraint. This kept transportation costs high and limited the gains from overland trading.

Coal was known in this period as a fuel for productive activities, but it was in limited supply; deep mining was as yet unfeasible owing to water inundating the mine galleries. High transportation costs were also limiting the widespread use of coal. We don't know what the price of coal was in this period, but where coal was used its price per unit of energy probably was close to that of wood.

The breakthrough came with the invention of the steam pump and the steam engine in England in the eighteenth century. The steam pump increased the supply of coal by making many more coal seams accessible. The steam engine made it possible to use the coal to generate kinetic energy, replacing expensive animal power with a cheaper alternative. The steam engine was the main driver of early industrialization. Railways, steamboats, and canals reduced transportation costs dramatically and allowed coal to be transported over long distances to the newly erected factories, where it fueled the steam engines that powered increasingly sophisticated machinery and equipment. The land needed to feed horses and other draft animals was no longer a limiting factor for economic growth. The world's industrialization was set in motion because a way had been found to tap the planet's carbon stocks—biomass and solar power collected and stored about 300 million years earlier—and make them useful for humanity, thereby liberating arable land for the production of food rather than fodder.

The inventions of the dynamo-electric motor and the internal-combustion engine in Germany in the second half of the nineteenth century gave industrialization a second boost by helping to make the new fossil energy available for ever more applications. Cheap kinetic energy now became available even for small machinery and small factories for which steam engines were too clumsy, and it became possible to manufacture cars and trucks and to build electric railways. Diesel-powered tractors replaced steam-powered tractors in agriculture in the early 1920s, further boosting agricultural productivity. Oil and gas discoveries complemented that development, providing the impulse for growth that wouldn't have been possible had mankind been restricted to bioenergy, which had hitherto placed it in the quandary of having to pay for growth with emptier stomachs.

Growth was amplified when agricultural productivity was boosted by the invention of chemical fertilizers by Justus von Liebig in the first half of the nineteenth century. Their synthetic production (based on the Haber-Bosch method, invented in the early years of the twentieth century) also required vast amounts of fossil energy.

The freeing of land for food production made unprecedented population growth possible. Figure 3.6 plots the world's population from about the end of the Roman period until today (and a UN projection to 2050). The kink in this curve at the time of the Industrial Revolution is obvious.

By the middle of the eighteenth century, when the Industrial Revolution began, the world's population was about 630 million. Today it is 6.9 billion. The average annual population growth rate since 1750 has thus been 0.9 percent. From the year 600 (when the world's population was 200 million) to 1750, the average annual population growth rate was just 0.1 percent.

Although the population grew exponentially after 1750, the world's economy grew even faster. Thanks to its deployment of carbon, mankind

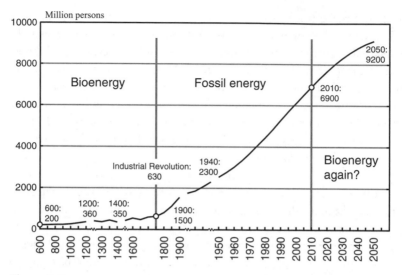

Figure 3.6
The path of the world's population. Sources: US Census Bureau, *Historical Estimates of World Population*, 2010; United Nations, *World Population Prospects: The 2008 Revision*, 2009.

managed, for the first time in history, to liberate itself from the Malthusian trap and raise the per-capita income of the masses above the subsistence level. Later, further advances in medicine and in birth control made it possible to secure those achievements in at least some parts of the world by slowing the rate of population growth, thereby ushering in a significant improvement in the standard of living.

The period in which food and energy markets were decoupled because of the cheapness of fossil carbon lasted a quarter of a millennium. It has now come to an end, as the Tortilla Crisis signaled. The increasing scarcity of fossil fuels, and to some extent the rising environmental concerns about depositing waste carbon in the atmosphere, have pushed the price of oil up to the level of the price of food, redirecting food from the table to the tank.

To summarize this story and give it a more abstract, theoretical interpretation, figure 3.7 illustrates mankind's development pattern in terms of carbon prices. Carbon prices are defined in terms of energy content and expressed relative to the wage of unskilled labor. The figure distinguishes three development periods and shows the time paths of the prices of biofuel and fossil fuel. The biofuel price is an average across various

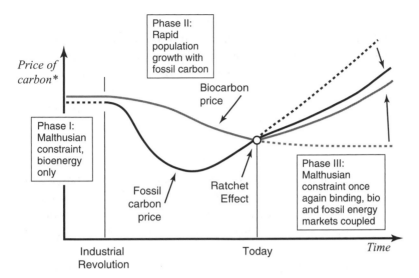

Figure 3.7
A brief history of fossil carbon and biocarbon. *Relative to unskilled wage, per unit of energy content.

kinds of food, fodder, and biofuels; the fossil-fuel price is an average of the prices of coal, crude oil, and gas.

In phase I, the markets for food and energy were closely interlinked, since draft and pack animals had to be nourished from the same land as humans. Coal was known, but it played no particular role in the economy. In the rare cases where coal was available and used in specific trades, its price must have been similar to that of bioenergy, in particular that of wood.

In phase II, steam pumps enabled deeper mining and increased the supply of coal, and steam engines made it possible to generate kinetic energy out of this coal in railways, steamships, and factories, triggering the Industrial Revolution. Later the other fossil fuels became available. Though fossil carbon couldn't directly be transformed into biocarbon, it gradually liberated biocarbon from its function as an energy resource, as the price of fossil carbon fell below the price of biocarbon. Land was converted from the production of fodder to food, and more food became available. The additional food set population growth in motion and allowed mankind to escape the Malthusian trap. Real wages for unskilled labor in terms of food increased after some delay, which is equivalent to saying that the price of biocarbon in terms of labor hours was falling, albeit by less than the price of fossil carbon. Real wages began to rise in the second half of the nineteenth century, and the rise accelerated in the twentieth century, when diesel-powered tractors and harvesting machines replaced draft animals in agriculture. The rising population and the growing capital accumulation that characterized this period eventually increased the demand for fossil carbon faster than the supply, pushing the price of fossil fuel up until it eventually (again) reached the price of biocarbon.

Phase III is starting now. The Ratchet Effect now links the two prices and the two carbon markets, inducing the sudden carbon leakage from the food market to the energy market that caused the Tortilla Crisis. Over time, more and more agricultural land will be devoted to the production of biofuels, and the conflict between energy and food will become more and more severe.

If the integration of agricultural land in the energy supply chain continues unchecked, the increase in the price of fossil carbon will slow down, and the reduced supply of biocarbon in the food market will accelerate the increases in the prices of food and biocarbon. If, in addition, policy makers impose constraints on the use of fossil carbon to fight global warming, or for security reasons, the rate of carbon leakage from the food market will accelerate, and food prices will rise even faster. This will put mankind back in the Malthusian trap, and many more people may starve to death. True, agriculture is much more productive today than it was at the time of the Industrial Revolution. Much less land is necessary to nourish a person than at that time. However, the additional productivity, including the portion attributable to synthetic fertilizers, stems largely from fossil carbon resources whose flow is to be reduced. And there are more than ten times as many people around now. To a large extent, these people owe their very existence to the exploitation of fossil carbon since the Industrial Revolution, and they can't be assumed away by wishful thinking simply because we now want to close the fossil carbon channel.

The development path described above is a horrific vision, and one hopes that it is not a realistic forecast. But if it were to prove true, the population forecast until 2050 depicted in figure 3.7 would not materialize. This would pose a problem not only for those who would starve but also for those who would have to defend themselves against the starving nations fighting for survival. If we decide to let this market mechanism work, and tolerate the replacement of fossil fuels with biofuel, and if we decide to continue to stimulate this replacement through policy measures with the avowed purpose of fighting global warming, hundreds of millions of people will be affected, and they will not merely take to the streets in peaceful protest. They will wage war.

Thus, the conclusion to be drawn from this chapter can only be that, with the exception of the use of biogenic waste, the biofuel option is a blind alley into which mankind should not venture. Instead of supporting biofuels, we should ban making such fuels from crops, rather than from biogenic waste, without delay. Now that the price of oil has reached the price of food, we should erect new barriers to prevent food from ending up in fuel tanks. Argentina, India, Russia, and other countries that banned the exporting of energy crops during the food crisis shouldn't be blamed for protecting their populations against a dangerous market development. Of course the problem of global warming is still there, but that problem is secondary to the horrors that a worldwide Tortilla Crisis would bring. The damage that food wars could cause around the world would dwarf the possible damages from global warming for the time being.

This doesn't mean that we should give up the fight against global warming, but it does mean that bioenergy simply isn't the right weapon for that.

Chapter 2 listed the technical options available to mankind in the form of "green" replacement technologies. There aren't many such options, alas. Wind power and photovoltaic devices, unfortunately, aren't serious alternatives to fossil power. However, nuclear fission and (perhaps) solar power projects such as Desertec provide alternatives to fossil fuels, if imperfect ones. Perhaps nuclear fusion will work someday, which would be a blessing for the poor and for those who otherwise would have to suffer from global warming. At present, however, mankind has no other option but to choose policies that would limit biofuel production, resource-intensive economic growth, and population growth.

to Hugo Chávez, the sheikhs, and all the other oil potentates, with a plea to leave more oil underground

4

The Neglected Supply Side

Reckoning without one's host

The climate problem is real and not an illusion. Because the damage caused by global warming could be immense, we have to act now. Some measures are indeed being undertaken. Consuming countries have capped their CO_2 emissions through the Kyoto Protocol. The European Union introduced an emissions allowance trading system. In addition, many countries have issued a large number of laws and regulations providing the necessary carrots and sticks. Every citizen tries to do his bit by reducing consumption. This concerted effort should bring some results.

Unfortunately, however, results are hard to detect. Indeed, results may not have been delivered at all. This chapter and the next discuss why that may have been the case.

Table 4.1 provides an overview of the concrete actions that have been undertaken as a result of all the laws, regulations, and incentives adopted with the aim of averting climate change by reducing consumption of fossil fuels and, with it, emissions of carbon dioxide.

To start with, we reduce the consumption of fossil fuels by becoming thriftier. We turn our central heating down a couple of degrees and put on a sweater, as Jimmy Carter once suggested. We turn the lights off when we leave a room, and we switch off the stand-by mode on our TV sets. With better insulation in our homes and refrigerators, we waste less energy. We drive less and use public transportation more, or we look for a job close to home.

We strive to devise intelligent measures to reduce fossil-fuel consumption—measures that simultaneously protect the environment and bring energy costs down. We replace incandescent light bulbs with

Table 4.1
What we are doing to curb CO_2 emissions.

Direct reduction of fossil fuel demand
Better insulation, lighter cars, less heating, less driving, switching off lights, switching off stand-by mode
More efficient use of energy
LEDs and fluorescent lights, variable-speed electric motors, intelligent energy management, common-rail diesel engines, DiesOtto engines, condensing boilers, gas-and-stream power plants, coal-fired plants burning finer powder coal, combined heat-and-power
"Green" electricity
Wind, hydro, solar, biomass, hybrid cars
Nuclear energy
Electricity and hydrogen generation
Other "green" energy sources
Pellets, wood chips, wood, biogas, biodiesel, bioethanol, heat pumps, solar thermopanels, geothermal energy

fluorescent or LED lamps, and we install automatic variable-speed circulating pumps in our heating systems in order to save electricity. Industry has increasingly turned to variable-speed electric motors for its machines and factories. Carmakers help by resorting to common-rail diesel engines. The new DiesOtto engine squeezes more energy out of the fuel than any other combustion motor. Condensing boilers are being favored for home installation. Power companies are turning to gas-and-steam power plants, which use the gas twice to drive the turbines. "Conventional" power plants grind their coal into ever-finer powder in order to leave no coal molecule unburned.

We are turning toward electricity from renewable sources—wind, water, sunlight, biomass. Wind turbines spin and solar panels glitter everywhere. Some cars store braking energy in batteries to use it again when power is needed to accelerate. Some people have power units in their basements that produce electricity and contribute their excess heat to the central heating.

Opinions are divided when it comes to nuclear power, because the final storage of its waste and the risks involved are variously assessed. Nuclear power is climate friendly inasmuch as it displaces fossil-fuel-generated electricity, and at present it offers the only realistic option for the hydrogen economy that many people dream of, though the Fukushima accident has dampened enthusiasm significantly.

We are now generating heat and electric power without using fossil fuels by burning wood, often in the form of pellets and wood chips. Our roofs are covered with solar thermopanels to heat water for our showers. We fill the tanks of our vehicles with biodiesel or with a gasoline-ethanol mix. Farmers heat their homes with gas generated from slurry and other waste. We pump heat out of the ground or the air. Some of us even bore deep into the ground to tap water heated by geothermal activity and use it to heat our homes.

Europe is particularly fond of "green" energy. France generates most of its electricity with nuclear reactors. Germany is the world champion in solar power and biodiesel, and ranks second in wind power after the much bigger United States. Sweden and Austria use a lot of hydropower, and Denmark's use of wind power is strikingly obvious when one flies over the country. Italy is covering its beautiful landscape with solar panels. In addition, all European countries impose high taxes on fossil fuels, and they have thousands of rules and regulations enforcing fuel efficiency, including a limit on the CO_2 output per kilometer of a car-maker's model fleet. The countries of eastern Europe have largely shut down the carbon-intensive factories of the communist years and have modernized their economies with more energy-efficient production techniques. As figure 4.1 shows, the result of all of this was an 11 percent reduction in CO_2 output relative to 1990, the Kyoto Protocol's base year. Germany, which together with Luxembourg accepted the largest percentage CO_2-reduction goal when the EU-Kyoto agreement was broken down to individual countries—a reduction of 21 percent from 1990 to the average of the years 2008–2012—has already surpassed that goal.

But do all these efforts help the climate? Can what the Europeans are doing really be expected to contribute to curbing global CO_2 emissions? Many insist that it depends on each individual, just as curbing the littering of roadsides does. Every individual can help keep the landscape pristine by deciding not to pollute it himself. The more individuals show discipline, the cleaner the roadsides. But that analogy fails on one decisive point: it doesn't consider that decisions to pollute or not to pollute the climate aren't independent of one another, but are linked directly to other such decisions through the global market for fossil fuels. What one country doesn't throw away will be thrown away by another. The CO_2 that we emit into the atmosphere came out of the ground as carbon, and we bought it on the world market for carbon. If the Germans buy and burn less coal, crude oil, or natural gas, the

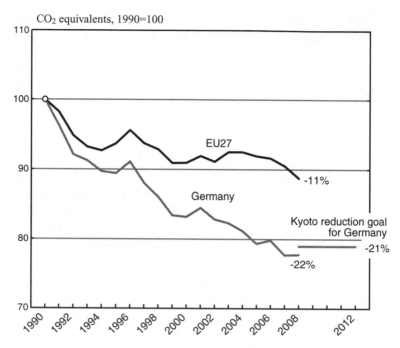

Figure 4.1
Europe's CO_2 emissions since 1990. Sources: European Environment Agency, *EEA Greenhouse Gas Data Viewer; Annual European Union Greenhouse Gas Inventory 1990–2008 and Inventory Report 2010,* EEA Technical Report 6/2010 (http://www.eea.europa.eu).

Chinese, say, will be able to buy and burn more. The analogy to littering would make sense only if one could assume that the European Union's reduction in consumption would translate into an exactly equivalent amount of carbon not being extracted by the oil sheikhs and other resource owners. But what if the resource owners remain utterly unimpressed by Europeans' efforts to curtail consumption and decide to extract, in every coming year, just as much crude oil and gas as they had planned to extract? The EU's sacrifices will then exert no influence whatsoever on overall CO_2 emissions, as all the oil and gas extracted will be burned somewhere. If the Europeans don't buy the fuels, someone else will buy them and burn them.

Anyone who believes that the European Union's reductions in CO_2 emissions will reduce global emissions by the same amount is doing his calculations without reckoning with the host. It isn't EU Commission

President Barroso, Chancellor Merkel, or the EU Parliament that sets emissions levels; it is oil and gas potentates such as Hugo Chávez, Mahmoud Ahmadinejad, Muammar Gaddafi, and even Vladimir Putin. It is they, together with the other owners of fossil-fuel resources, who will ultimately determine how much carbon will be extracted and burned, eventually reaching the atmosphere as CO_2.

For this reason, a rational climate policy must involve the resource owners, and it must develop strategies that nudge them toward a more conservative extraction behavior. There isn't the slightest glimmer of this in the policies listed in table 4.1. All those measures assume that the resource suppliers will meekly take directions from our fuel purchases, and that they have no agenda of their own. In other words, the measures assume that we buyers will determine how much crude oil will be pumped out of the ground, and that the suppliers will have to adapt to our wishes. That is the implicit assumption of the policy makers. That it is more than a bit naive goes without saying.

The doubts are reinforced when we consider the evolution of global CO_2 emissions, illustrated in figure 4.2. The curve exhibits not the slightest indication that the consumption reduction in Germany and other European countries, which gathered pace with the Kyoto Protocol commitments, has exerted any influence on worldwide greenhouse-gas

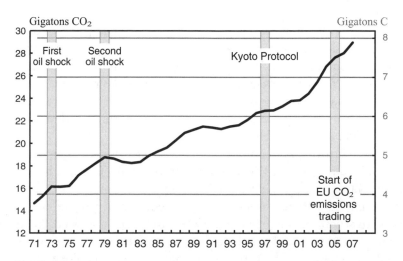

Figure 4.2
The world's CO_2 emissions. Sources: International Energy Agency, Database, *CO_2 Emissions from Fuel Combustion* (http://www.sourceoecd.org).

emissions. On the contrary, the curve appears to suggest that recently, since the adoption of the Kyoto Protocol, they have accelerated.

The missing regulating screw

But are the data plotted in figure 4.2 reliable? After all, it is difficult to measure CO_2 emissions directly. Usually there are no metering devices or sensors attached to chimneys, vehicle exhausts, or smokestacks. The actual method normally used is indirect: it measures fossil-fuel consumption, then extrapolates the amount of CO_2 emitted. Can it be that we overlook the efforts made by the Western world to increase the efficiency of its motors and combustion processes, which would then reduce the amount of CO_2 emitted per unit of fuel? Are there not formidable possibilities to reduce CO_2 emissions from any specified amount of fossil fuel? Do we not perhaps have scope for decisions beyond that accorded us by the resource owners?

Unfortunately, the answer is negative. Leaving aside sequestration, which was discussed in chapter 2 in association with the storage of nuclear waste, there simply is no scope. The reason for this is chemistry. The carbon atoms that are released into the atmosphere when fossil fuels are burned are the same atoms that the owners of fossil-fuel resources extracted from deep underground. Nothing is added, nothing subtracted. The amounts are identical.

Let us resume the discussion begun in chapter 1, paying closer attention to coal, crude oil, and natural gas. All three are hydrocarbons—that is, chemical compounds of carbon and hydrogen. Both carbon and hydrogen occur in such fuels largely in a non-oxidized, "reduced" state. In order for them to produce useful energy, they must react with oxygen, which is what happens in combustion. Combustion turns carbon into carbon dioxide, and hydrogen into water.

Coal contains a very low proportion of hydrogen: about 3–6 percent.[1] Crude oil contains between five and nine carbon atoms for each hydrogen atom. Natural gas, or methane, binds four hydrogen atoms to each carbon atom. A hydrogen atom generates about 30.7 percent of the energy of a carbon atom upon combustion.[2] A methane molecule, made up of one carbon atom and four hydrogen atoms, delivers through combustion 2.23 times as much energy as the one carbon atom alone. If the binding energy is deducted, you still get 2.04 times as much. Thus, natural gas generates a bit more than twice as much energy for the same amount of carbon dioxide emissions as pure carbon. That is why natural gas has such good environmental credentials and coal such poor ones. A ton of methane delivers 65 percent more energy than a ton of anthracite, and 3 percent less CO_2. And a car powered by natural gas is more environmentally friendly than one powered by gasoline because it runs a bit more than half of the total distance on hydrogen.

Although the ratio of energy delivered to CO_2 emissions is much better for methane than for crude oil, and the ratio for crude oil is better than that for coal, all these fuels emit carbon dioxide upon combustion. One might imagine chemists or engineers pulling a couple of tools from their bag of tricks to make it possible to keep the amount of energy delivered unaltered while reducing the associated emission of carbon dioxide, but that isn't going to happen. The number of CO_2 molecules released is, for each of the three fossil fuels in question, strictly proportional to the amount of energy delivered, with a different ratio for each of the fuels, and it is identical to the number of carbon atoms burned.

It is, of course, true that by using more efficient motors and combustion processes we can reduce the amount of fossil fuel needed to perform a certain amount of work. The car industry, for instance, has made great strides in reducing the burning of fuel by more efficiently turning waste heat into kinetic energy. This does improve motor efficiency, reducing the amount of fuel needed, but doesn't alter the fact that, even in this efficient process, all the carbon burned is turned into carbon dioxide.

A small qualification must be made for the fact that efficiency also has been improved by burning an ever-larger proportion of the carbon in the fuel, instead of letting it escape from exhausts or smokestacks as soot. Just think of the coal-fired power plants that grind their coal to ever-finer powder, or the common-rail diesel motors that perform up to five fuel injections per stroke, maximizing combustion of the fuel delivered. Thick black soot clouds emanating from smokestacks and vehicle exhausts are now a thing of the past. Soot consists of unburned coal particles. The residual energy they contain used to be wasted, but the new processes allow us to tap it. With modern motors and furnaces, the energy loss in soot is in the range of tenths of a percent. However, such efficiency improvements haven't reduced the amount of CO_2 emitted, as many people tend to believe, but instead have increased it. More energy is now obtained out of each ton of crude oil or coal extracted from the ground by having an ever-higher proportion of the carbon atoms joined through combustion with two oxygen atoms than would have been obtained in the absence of such technical improvements.

But what does that imply for the climate? Nothing. We would have to roll back all the technical achievements and go back to the old sooty combustion if we wanted to reduce the amount of CO_2 resulting from each ton of carbon combusted. The possibilities for efficiency gains of this kind are largely exhausted anyway, so one can now equate the amount of carbon entering the combustion process with the amount of carbon actually burned and released into the atmosphere.

Apart from CO_2 sequestration, there is no leeway, and no regulating screw that we could tighten to decouple CO_2 emissions from purchases of carbon in the world market for fossil fuels. For that reason, it makes no sense to orient climate policies exclusively toward the methods listed in table 4.1 without also developing a strategy to convince the oil sheikhs, the American coal companies, and the owners of the Siberian oil fields to leave more of their resources underground.

Supply and demand

From an economic perspective, the issue here is supply and demand in the world market for fossil fuels. Demand is the amount that we and other consumers would like to buy; supply is the amount that the resource owners want to sell. Both supply and demand express wishes. Whether those wishes are realized depends on whether the representatives of the two ends of the market can reach an agreement and close a contract. The wishes are usually but not always realized.

All the measures described in table 4.1— the measures taken in the European Union and in other parts of the world to curb carbon dioxide emissions—aim at reducing demand for fossil fuels. Not one of them could even be interpreted as trying to influence those who are responsible for the supply side. The supply side has been thoroughly neglected. It doesn't crop up in public debates or in the design of "green" policies. The wish list is diligently compiled, but it isn't delivered.

Astonishingly, supply doesn't figure much in the climate-related scientific literature either. Though everyone knows that carbon dioxide is emitted through combustion of fossil fuels, climate studies that make supply decisions in the global market an integral part of their analyses are rare.[3] Either the energy issue is discussed (in which case resource stocks and reserves-to-production ratios occupy center stage) or the climate-change problem is discussed (in which case the focus is on meteorological and engineering issues, centering mostly on the technical possibilities for curbing CO_2 emissions). Even in the comprehensive Stern Review, which reopened the climate debate in Europe in 2006, the supply issue is barely broached.[4] The Stern Review does examine many economic aspects of the problem, in particular when comparing the damage wrought by climate change with the mitigation costs, yet only one or two of its 600 pages contain remarks on supply issues.

This chapter attempts to close that gap. In order to keep the analysis as simple as possible, let us abstract for a while from the fact that the carbon market is divided into many submarkets for different kinds of fuels with different characteristics and prices. Carbon fuels can substitute pretty well for each other, and chemical processes even make it possible to transform them into each other. The fact that some fossil fuels contain substantial portions of hydrogen, or the fact that some are easier to transport because they are in liquid form, can be captured with a surcharge above the basic price of carbon, akin to the difference between the base price of a car and the price of its options. Because the hydrogen content of carbon fuels is of no further interest for climate considerations, let us concentrate on the carbon content.

In order to understand the problem fully, and in particular in order to better devise an effective climate policy, it is necessary to understand the supply decisions of the resource owners. This isn't easy. Unlike other goods, for which supply decisions affect only the present and the immediate future, decisions concerning the supply of natural resources involve extremely long time horizons that include considerations aimed at sensible exploitation of the resource until its depletion. In other words, these are decisions whose implications will affect the current owners and their heirs over many decades, perhaps centuries.

But let us start at the beginning. Before we tackle such long perspectives, we should make matters easier by performing a supply-and-demand analysis over a single period. In order to avoid getting entangled in short-run economic issues, such as fluctuations in the business cycle, let's assume that the period is 10 years.

Supply and demand are brought into equilibrium by prices. A high price leads to excessive supply, a low price to excessive demand. Somewhere in the middle there is a price that brings the market to equilibrium. A higher price will not prevail, because the suppliers will underbid one another and push the price down again. A lower price will not prevail,

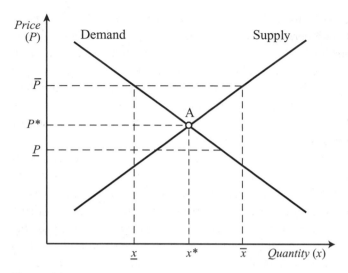

Figure 4.3
Supply and demand.

because the buyers will outbid one another and thus push the price up again. Only a price at which just as much is demanded as supplied can endure. This price brings the wishes of both sides of the market into agreement and determines the amount traded in the market.

In a well-functioning competitive market, the equilibrium price is approximately the same for all sellers and all buyers. That is the law of one price, which was discussed in chapter 2 in the context of the European cap-and-trade system. There the price referred to emission certificates; here it refers to the world market price of fossil carbon.

Figure 4.3 illustrates this relationship with a diagram that can be found in every first-semester economics textbook. Those who are familiar with this diagram can move on to the next section. Those who aren't familiar with it should devote a moment to it, because a sound understanding of supply and demand is essential for what comes next.

Figure 4.3 shows how much both sides of the market are willing to trade at a certain price. In the simplest case, supply and demand can be depicted as straight lines, as in this figure; at this point that isn't relevant, however. The demand curve shows the sum of the purchasing desires of all consumers, while the supply curve shows the selling desires of all producers. To read the supply and demand values, one moves horizontally from the price axis to the curve in question and then vertically down to the quantity axis; here the figure differs from what you may remember from your mathematics classes. P and x stand for price and quantity. If, for example, you move from price $\overline{P}$ on the price axis along the dashed line to the right in order to see what this price means for supply and demand, you will intersect the demand curve and, by going straight down to the quantity axis, read the corresponding quantity $\underline{x}$ that represents the willingness to trade on the part of the consumers. (A scale with real values isn't necessary here; we are merely dealing with the principle.) Farther to the right, the dashed line intersects the supply curve, which leads downward to the value $\overline{x}$ on the quantity scale; $\overline{x}$ stands for the trade desires of the producers. The two desires evidently don't yet coincide at price $\overline{P}$.

The demand curve slopes downward because a lower price prompts consumers to buy more; the supply curve slopes upward because at a higher price producers want to sell more. Behind this is the idea that consumers will switch from one good to another if the latter is cheaper, and that more producers will be able to stay in the market if the price is higher, or that existing producers, when the price is high enough, will build more factories to produce and sell more of the good in question.

Given that at price $\overline{P}$ supply exceeds demand by $\overline{x} - \underline{x}$, this price doesn't represent a market equilibrium. Producers will make price concessions to the consumers, as otherwise they will not be able to sell their wares, while the consumers will take advantage of the situation by trying to get the producers to lower the prices even further. Only at point A, where the two curves intersect and to which price P^* and quantity x^* correspond, are the trade desires of both sides of the market compatible. This is the point of equilibrium. When the price lies below this level, for instance at $\underline{P}$, demand exceeds supply, so the price will move upward. The producers take advantage of the situation and demand a higher price, and the consumers pay this price in order to get the good at all. As the price rises, the consumers become less willing to buy the product, but the producers are willing to sell more. The price stops rising when the equilibrium price, P^*, is reached. At that price, trade occurs and a quantity x^* changes hands.

Because supply and demand together determine the quantity traded, both must as a rule move in order to find the price at which they are compatible and the market settles in its equilibrium. For this reason, these two values cannot be observed separately. The quantity traded doesn't show what the wishes were before they were brought together by the price mechanism. Occasionally one sees in the press a graph captioned "Crude oil demand," or an assertion that "the world's energy demand has doubled in x years." Though not wrong, these are misleading, because instead of "demand" one could just as easily say "supply." Statistics rarely show demand or supply; they just show the volume of transactions resulting from both.

Supply and demand can be observed separately only in markets with regulated prices, which don't deliver a market equilibrium. For example, the supply of wind-generated electricity can be inferred from the amount of such electricity delivered if the state pays a feed-in remuneration that lies above the equilibrium price and, on top of that, has mandated that the power utilities take up that kind of electricity. Our graph would depict that situation with a price set at $\overline{P}$ and a quantity at $\overline{x}$. This quantity represents only the supply, not a demand that would correspond to a voluntary purchasing desire on the part of the power utilities, such as $\underline{x}$ in the figure. Another example is a labor market with binding official wages. In that case, the demand for labor on the part of the employers can be inferred from their actual employment of personnel, but the employment cannot be equated with the supply by households. This case can be captured by our graph if $\overline{P}$ is interpreted as a binding wage and $\underline{x}$ as the number of people employed and demanded. Given that with wages set at this level $\overline{x}$ workers would be willing to offer their services, a supply excess on the order of $\overline{x} - \underline{x}$ will occur, which is otherwise known as unemployment.

How "green" policies shift the demand curve

A country's demand for carbon depends on many things, including its gross domestic product, its population, and its climate. It is also driven by price. For this reason, we can depict the demand curve for carbon as in figure 4.3. The fact that demand decreases with rising prices results in part from individuals' cutting back on consumption—that is, abstaining from energy-intensive undertakings such as vacations in faraway places and jaunts by car to the countryside. However, this is only one aspect. Perhaps even more important is the fact that with a rising price alternative technologies come into play. The higher the price of fossil carbon, the more profitable the renewable forms of obtaining energy, such as those listed in table 4.1. Processes already known are implemented, and companies put their engineers to work devising new ways to use energy more efficiently. All of this reduces demand for fossil carbon. Thus, to a large extent, a demand curve depicts replacement or "backstop" technologies becoming profitable at alternative prices.

A good example of the importance of prices for the implementation of alternative technologies is a compact car from Volkswagen known as the 3-Liter Lupo. The moniker had to do with the Lupo's purported ability to travel 100 kilometers on 3 liters of fuel—that is, about 78 miles per US gallon. Conceived in the 1980s under the influence of the last oil crisis, and developed in the 1990s, the Lupo was put into production in 1998. But when it debuted, the price of oil was low, and it flopped in the market. Ten years later, oil prices were once again high, so Volkswagen gave the Lupo another try. In 2010 it was reintroduced.

If the price of oil were as low as it was 40 years ago, we would not be so keen on insulating our houses as we are now, we would be content to use low-efficiency heating systems, and we would continue to drive gas guzzlers. There would surely have been technical progress, but it would probably have gone toward increasing horsepower, with concomitantly higher fuel consumption. Who knows, perhaps Volkswagen would have produced a Rabbit with a V8 engine. Examples in this direction aren't exactly rare. Our descendants will shake their heads when someone shows them pictures of compact cars equipped with monster engines from the days of cheap, abundant energy.

Figure 4.4 shows two demand curves for fossil fuels. The upper curve corresponds to a case in which there is no government intervention, the lower curve to a case in which there is a government environmental policy. These two curves describe hypothetical scenarios for the present decade. The demand curves for later decades would be similar, but with different positioning of the curves.

Let us first consider the upper curve. It illustrates the price-driven behavior of people and firms when making technology choices without government intervention in the sense of environmental prescriptions, feed-in tariffs, subsidies, and the like. The price is the world market price of carbon, which hypothetically can have different values. Four points along this curve have been chosen as examples: A, B, C, and D. Point A represents wasteful use of energy, such as gas-guzzling cars and poorly insulated homes. This point would have been reached today if oil prices had remained as low as they were in the past. At a somewhat higher price (point B), common-rail diesel engines have become widely accepted,

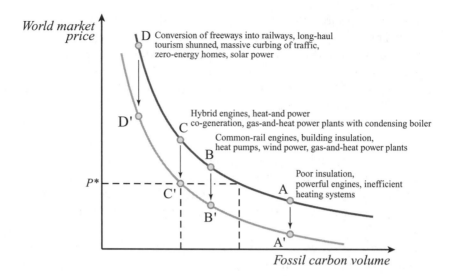

Figure 4.4
Possible price-quality scenarios: two alternative positions for the demand curve in the present. The upper curve corresponds to a pure market scenario without any sort of government policy interfering with private decisions. The lower curve corresponds to a scenario with "green" policies in place. Each point on the upper curve is associated with a certain pattern of energy using technologies and behavioral attitudes of households and firms. At a sufficiently low price (point A), people would continue to insulate their houses poorly and cars would be equipped with gas-guzzling engines, as in the recent past. At a sufficiently high price (point D), in contrast, people would be building "zero-energy houses," shunning long-distance holidays, and using solar power on a large scale—something that at present is only expected for some future time. The points lying straight downward on the lower curve give the prices that would correspond to these same behavior and technology patterns if "green" policies were pursued. They are all lower than those on the upper curve because "green" policy incentives imply that the corresponding energy saving strategies already pay off at lower world market prices. All hypothetical price-quantity combinations shown on the graph refer to the same point in time, say a given decade. There are similar diagrams for all future decades, only with different positioning of the two curves, as the other factors influencing demand that aren't captured in prices (domestic product, climate, population level, etc.) would also be likely to change.

gas-and-steam power plants are the norm, and some heating is being provided by heat pumps. Fossil-fuel consumption is correspondingly lower than at point A. At a still higher price (point C), hybrid cars and cogeneration of heat and power have become normal. The gas-and-steam power plants have even been extended by condensing boilers to collect the last bit of energy from the exhaust. At point D, the price of fuel is so high that driving a car wouldn't be an option for most people, and no fossil fuels would be burned for heating. People would be living in "zero-energy homes" (that is, houses with zero net energy consumption and zero carbon emissions), and half of the freeways would have been converted to railways. Because long flights would be prohibitively expensive, most vacations would be taken closer to home.

Point D is still something for some future time, because the price of oil isn't as high today as it would be in such a hypothetical scenario. But it would indeed describe the present if the price were already at the level we can expect for the future. Older readers will remember how such a scenario for the times in which we are now living was described 30 years ago. And because things turned out otherwise and the price of oil didn't rise at all in the period 1980–2004, what then was expected for today continues to be something for the future.

The demand curve in figure 4.4, unlike that in 4.3, isn't depicted as a straight line, but as a concave one that bends upward with increasing prices. The reason for the bending is that the residual demand will probably never disappear completely, no matter how high the price may climb, since there isn't a substitute for every single use of fossil fuels.

As an example, figure 4.4 assumes that the world market price of fossil carbon is P^*. At that price, the technologies represented by point A are no longer chosen. Gas guzzlers are frowned upon, and houses and apartments lacking insulation find no takers. On the other hand, the price isn't yet high enough to make common-rail diesel engines, heat pumps, or biofuels generally attractive. Demand is lower than in the case of point A, but not as low as it would be if all these technical alternatives were used. Evidently that isn't the scenario that we observe today. However, the demand curve doesn't show what is happening today; rather, it shows what *would* happen today at alternative prices without government interference in the decisions of people. These are purely hypothetical scenarios. We need a bit of counterfactual history in order to understand the present.[5]

In general, the position of the demand curve depends on many influencing factors, among them state policies. The lower curve shows the demand for carbon as a function of price for the "green" policies currently in force. Let us assume that these policies are effective and that they will not just go up in smoke under the mechanisms of emissions trading, as was discussed in chapter 2. Whether one thinks of environmental taxation, feed-in tariffs, or the mandated blending of biofuel, the state strives to grant "green" technologies a competitive advantage and help them achieve an earlier breakthrough than what a rising fuel price itself would imply. That means that the price thresholds at which the corresponding "green" technologies would become competitive are all lower than without the green policies. Figure 4.4 shows this graphically by having all the points on the demand curve move down, from A, B, C, and D to A', B', C', and D'.

If the price thresholds for the competitiveness of "green" technologies fall, that doesn't necessarily mean that the price of fossil fuel will decrease. Suppose, for example, that P^* is a firm world price (say, because the policy measures are taken by only one country, and that country is too small to affect the world market price of carbon significantly). Demand then contracts and is represented as point C' on the demand curve. Point C' corresponds to point C on the original (upper) demand curve, which corresponds to the energy-saving efforts as described in the figure. The freeways are still busy, and occasionally people still indulge in flights to far-away vacation paradises.

Rembrandts vs. cars: The carbon supply

Let us now consider supply. How strongly supply co-determines the quantity of goods depends on whether supply is rigid or price-elastic. In the market for paintings by Rembrandt, the supply side has the better cards, because Rembrandt is dead and cannot paint anymore. The price is determined by how much the demand side is willing to spend, and the quantity of Rembrandts is determined entirely by the supply side. Note that economists define demand as including the "own demand" of the happy owners of Rembrandts. The same is true of plots of land that are strictly limited in number in certain urban locations. In this case, too, the supply is rigid or inelastic and therefore determines the quantity entirely; conversely, for that reason, demand is *relatively* elastic, because not everyone can afford a plot of land in the best part of town at any given price.

The picture is different in the car market. Since cars can be produced in any number, carmakers produce as many cars as consumers are willing to buy. If demand increases, a short-term supply shortfall may occur, pushing prices up somewhat. But the carmakers will then mandate extra work shifts or perhaps even expand production capacity so that production grows along with demand. And if demand contracts, a price war may erupt, bringing the prices down and making the buyers happy, because producers don't want to get stuck with vehicles they have already made. Soon, however, some factories will be shut down until the prices return to normal. The normal price of a car is determined by the production costs of cars of a certain quality and size, and in the middle or the long term it doesn't depend particularly strongly on the demand. A price that is a bit higher than the break-even price already prompts carmakers to expand their production significantly, and no cars will be offered if the price falls just a bit below the break-even point. That is what is meant by elastic supply: small price changes cause large changes in supply.

Figure 4.5 illustrates these two alternatives. A reduction in demand (that is, a leftward shift of the demand curve) shifts the equilibrium from A to B in both panels. In the left panel, supply is elastic, and the shift is a quantity reduction; in the right panel, supply is inelastic, and the shift is a price reduction.

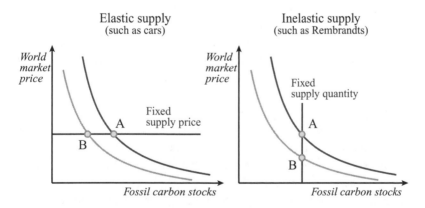

Figure 4.5
The two extreme cases of perfectly elastic supply (left) and perfectly inelastic supply (right). These graphs apply to a given period (decade). As will be explained in chapter 5, the first case applies more to temporary and the second more to permanent changes in the demand for fossil fuels.

Environmental policy makers assume, evidently, that the supply of carbon is elastic, as in the case of cars—that demand drives the supply of carbon. They assume that reductions in demand resulting from their policy measures translate fully into corresponding reductions in the quantity of carbon extracted, because the producers have firm notions about the price of their products and will not budge from those notions even if demand decreases. But if supply is rigid, as depicted in the right-hand panel of figure 4.5, "green" policies will simply trigger a decrease in the world price of carbon from A to B, and the quantity sold and consumed will stay unchanged. In this case, the price must decrease enough to induce additional demand equal to the demand lost at the original price as a result of the application of "green" policies. Taxes, high feed-in tariffs, and mandated biofuel blends reduce demand for carbon at any world price, but the resulting excessive supply brings prices down, and that, in turn, stimulates demand until the same amount of carbon is consumed as before. "Green" policies are, in this case, ineffectual in reducing the output of carbon dioxide.

One could argue that in the case of small countries only the elastic type of supply would be relevant. It cannot be assumed, after all, that a single country's carbon purchases would influence the carbon price in the world market in any meaningful measure. This argument is not at all wrong. If what we mean by "supply" is the quantity the world markets make available to the single country, then this supply is indeed elastic, which only means that we have to accept the world market price as it is. If we pay more, the entire world will be keen on selling to us; if we pay less, nobody will want to sell us anything. However, the supply curves shown in figure 4.5 don't refer to the supply available to a single country; they refer to the supply to the whole world—that is, the quantity that all the producers in the world bring to market. Likewise, the demand curves don't represent a single country alone; they represent all consumer countries combined. But in the face of rigid supply, as shown on the right-hand side of figure 4.5, not even acting together can the consumer countries influence the world's climate.

The promotion of alternative energy sources such as wind and sunlight doesn't slow down climate change if the carbon supply is rigid; instead, it increases the world's energy consumption by the amount brought about by these alternative sources, as the oil and gas wells continue producing and their owners reduce their prices so far as to give rise to a demand that, regardless of "green" policies, is just as high as it was before. For that same reason, nuclear power as such doesn't help the climate either if the supply of fossil carbon is fixed. Just as with the "green" energies, nuclear energy adds itself to the fossil variety offered by the resource owners, instead of displacing it, because it depresses the carbon price sufficiently. The production of biodiesel, ethanol, and pellets is likewise just neutral, because it releases into the atmosphere carbon that photosynthesis had previously removed from it. No positive effects resulting from a displacement of fossil fuels can be observed if supply is price-inelastic.

Of course, all of this is not to say that supply is really inelastic. It just shows that it isn't permissible to neglect the supply side. In fact, the carbon supply isn't a simple function of the current price, neither elastic nor inelastic; it depends on a time sequence of expected prices in all periods of time, from the present on to infinity. Before we discuss that, let us see what the assumption of a rigid world supply implies for the distribution of carbon consumption across the world.

Carbon leakage: Grabbing from the collection box

With a rigid world supply, the reduction in the world market price induces as much additional demand as is lost as a result of pursuing "green" policies at the original world market price. But while demand reductions induced by "green" policy measures occur only in some countries, the resulting lower world market price stimulates demand in all countries, whether or not they pursue "green" policies. Because the quantity demanded by all countries taken together is determined by a supply that is rigid, the carbon quantities relinquished by the "green" countries are taken up by other countries. Although demand in the "green" countries will also increase after a fall in the world market price relative to the decrease caused at the old price by the "green" policy measures, it cannot increase back to where it was before, because if it did the other countries wouldn't be able to increase their consumption as prices fall, as they are doing.[6]

That demand increases in the other countries may come as something of a surprise. It is therefore useful to study these relationships more closely. Figure 4.6 helps in this regard. It divides the world into two groups of consumers. One group consists of countries that have accepted commitments based on the Kyoto Protocol and that pursue a wide range of policies aimed at reducing their demand for fossil fuels relative to what

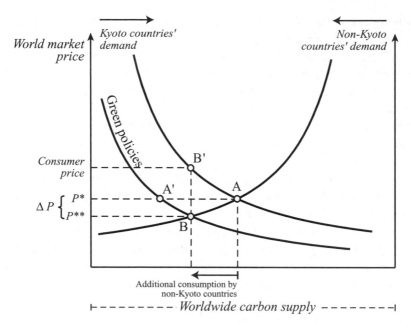

Figure 4.6
Consumer competition for carbon: how Europe subsidizes the world's carbon consumption. This graph refers to market behavior in a given period (decade). It illustrates how demand-reducing measures by a group of countries can affect the apportionment of demand among various groups of countries *at a given level of worldwide supply.* Supply is measured by the horizontal distance between the vertical axes, that is, by the dashed line at the bottom. (The assumption of a given level of supply will be dispensed with in chapter 5. That chapter will show that the changes in ΔP between the periods play a decisive role in how supply is apportioned among the individual periods.)

the case would be in absence of such commitments. This group includes the EU 27, Japan, Russia, Ukraine, Canada, and Australia. The other group consists of all the other countries in the world, including the United States, India, China, Brazil, Korea, Indonesia, Mexico, and Iran, that have signed the Kyoto Protocol but are not subject to any commitments to restrict their emissions. The countries that have accepted emissions ceilings will be referred to as "Kyoto countries" and the rest as "non-Kyoto countries."

The distance between the two vertical axes in figure 4.6 is the supply of the resource owners. This supply is assumed to be price-inelastic, an assumption that will later be replaced with an intertemporal supply hypothesis. The figure shows the demand of the Kyoto countries from left to right, and the demand of the non-Kyoto countries from right to left. The upper of the two downward-sloping curves is the demand curve of the Kyoto countries when no "green" policies are pursued. The lower curve is their corresponding demand curve with such policies. The upward-sloping curve shows the demand of the non-Kyoto countries. It is similar to the demand curve of the Kyoto countries, but it is a mirror image. If read from right to left, it is downward-sloping, too. There is only one variant of this curve, as it is assumed that only the Kyoto countries take active policy measures against carbon demand.

The intersection point of the demand curves (A) represents the market equilibrium on the world market for carbon in the basic scenario in which no "green" policies are pursued, since in this case the sum of the quantities demanded by both groups of countries equals the supply (represented by the width of the diagram). The quantity that the Kyoto countries purchase is represented by the distance from point A to the left vertical line of the diagram; the quantity that the non-Kyoto countries purchase is represented by the distance from point A to the right vertical line.

When the Kyoto countries implement their "green" policies, their demand curve shifts from the upper to the lower position. Relative to the base scenario, the demand at price P^* would shrink from A to A', and hence P^* no longer represents an equilibrium price. Excessive supply lets the price fall until both groups of countries buy sufficiently more to again absorb the supply. The price reduction in the diagram is ΔP, and the new price is P^{**}. The price reduction increases demand again somewhat in the Kyoto countries; however, this cannot compensate for the reduction in demand due to the "green" policy, because the non-Kyoto countries also buy more when the price is lower. The new equilibrium occurs at point B. It describes the balance between supply and demand when some countries follow "green" policies and the others don't. The carbon not burned by the Kyoto countries just goes to the non-Kyoto countries and is burned there, as represented by the arrow below the diagram. This result, which follows logically from the assumption of a given supply, demonstrates once again how important it is to think about the forces that could determine supply.

The non-Kyoto countries, which account for about 70 percent of global CO_2 emissions, are the beneficiaries of this policy. Thanks to the voluntary restraint of the Kyoto countries, they can purchase their carbon more cheaply and therefore consume more of it. This makes it possible for Americans to drive huge cars and SUVs, and for the growth- and energy-hungry Chinese to make no particular effort to change their polluting economic ways. The Europeans subsidize American and Chinese consumers without doing anything for the climate in the process.

Consumers in the Kyoto countries get nothing out of their demand restrictions. On the contrary, they have to foot the bill of a useless environmental policy. Because less carbon is available in their country, they pay a higher consumer price, represented by point B'. This can best be understood if we assume that the Kyoto countries impose a tax on the consumption of fossil fuels. This tax drives a wedge between the world market price and the consumer price, represented by the distance B'B in the diagram. Because domestic consumers pay more, they buy less; because they buy less, the world market price falls. Similar effects would occur if a cap-and-trade system were to be used to restrain demand in the Kyoto countries.

Note, however, that there is also a positive side effect for the Kyoto countries, insofar as their demand restraint implies that *in toto* they buy the fossil fuels more cheaply than without such a restraint. In the case of a tax or a sale of emission allowances, the additional revenue of the government must also be taken into account, and in the case of free allocation of emission rights an increased profit margin for polluters. Thus, there is a rationale for such policies from the viewpoint of the Kyoto countries, since they might try to generate what economists call a *monopsony profit*—that is, a profit from creating a demand cartel. However, such measures are of no help in the fight against global warming unless they are able to reduce the supply of fossil energy. Moreover, if the Kyoto countries reduce their demand not by means of taxation or a cap-and-trade system but by forcing on their citizens more expensive ways of producing electricity (perhaps by subsidizing "green" technologies via feed-in tariffs), there isn't even a positive side effect such as higher government revenues or larger profit margins; there is only a waste of resources in the Kyoto countries that makes them poorer but benefits consumers in the non-Kyoto countries.

Some think that the efforts to curb climate change are like something that might happen in a church: The Kyoto countries, as they leave the church, drop some money in the collection box. Behind them come the non-Kyoto countries. The former know that the latter are stingy and don't want to contribute any money. They hope, however, that their good example will encourage the others to contribute. And if the non-Kyoto countries don't contribute anything, at least the donation has been for a good cause.

But the church metaphor is wrong on two counts. For one thing, the result isn't correct. Not only do the non-Kyoto countries decline to put money in the box; they grab from it the money that the Kyoto counties deposited. Nothing is left for the good cause. For another, when it comes to the climate, the force behind contributing isn't morality or philosophy, both of which play a big role in contributing money to a church's collection box. The real forces are the hard laws that rule the world market for resources. Consumers in the non-Kyoto countries simply don't see that the fossil-fuel prices, which are making them groan, are lower than they would otherwise be thanks to the efforts of the Kyoto countries— and they don't really care about that. They just buy what they need and can afford, without asking themselves any moral questions.[7]

If the supply of fossil fuels is rigid, it is inevitable that the savings efforts of individual countries will be negated by the countervailing actions of other countries. This happens regardless of whether the countries are also bound to each other by the capital or product markets. The stronger such links, however, the more elastic the consumption reaction of all countries to price movements, and the smaller the price decreases triggered by the changes in carbon quantities demanded. For this reason, public and scientific discussions have focused some attention on the probable displacement of CO_2-intensive industries from the Kyoto countries to the non-Kyoto countries, from which they supply their wares to the world markets.[8] The term that characterizes this discussion is "carbon leakage." This is undoubtedly an important issue. With fierce competition in the world's product markets, under certain circumstances a minor increase in production costs (for instance, one caused by an eco-tax) can suffice to nudge European producers out of the market, clearing the way for their Chinese competitors. For this reason, President Nicolas Sarkozy of France pleads for EU protection against the importing of CO_2-intensive goods from non-Kyoto countries.[9] That is an understandable demand. But even if Sarkozy were to get his way and so prevent the exodus of certain industries, it would not change the fact that the carbon extracted and sold by the resource owners will eventually be burned somewhere in the world. It isn't at all necessary to assume that jobs can move by way of capital flows or changes in the international division of labor. Even if no country had a trade link for produced goods with another one and none were connected to others via the capital market, carbon leakage would self-evidently occur via the market for carbon itself.

When it comes to saving the planet instead of saving jobs, punitive tariffs are no help if supply is as rigid as has heretofore been assumed. Apart from sequestration, this remains an unavoidable truth: if and only if a way is found to reduce the carbon supply will environmental policies be able to mitigate global warming.

Whether supply is truly rigid is quite a different question. Nothing has been said yet about what actually determines supply. In the long run, supply is certainly rigid: nature deposited a certain amount of carbon during the Carboniferous and no more. Figure 4.6, however, doesn't refer to the long run, but only to one period, and here supply could well be increased or reduced, to the benefit or the detriment of other periods.

Nature's supply

Let us now approach the intertemporal dimension of the supply problem by taking a look at the quantities of natural resources that nature has to offer. These quantities are the basis for the supplies the resource owners bring to market in each period.

The supply that the owners of fossil carbon bring to the market in a period must first be extracted from the stocks they own. Some mines and deposits are already depleted; others are about to be depleted; others are yet to be tapped (either because their owners want to keep them in reserve or because they lie in such inconvenient locations that their exploitation will make economic sense only when prices are higher). Some stocks haven't been discovered yet, but these are becoming increasingly rare; the number of new discoveries has diminished steadily over the years. When it comes to resource stocks today, the picture is similar to what happened with world maps at the end of the seventeenth century. By that time, maps contained very few blank areas, because sailors had already explored nearly the entire world. But exceptions, of course, prove the rule. Only a short while ago, a big discovery was reported in Brazil that would be equal to about 2 percent of the known conventional oil reserves.[10]

Not long ago, it was feared that the world's reserves of fossil fuels would be soon depleted. That topic was covered in the context of the running times of resource stocks in chapter 2—see, in particular, figure 2.6. After 1972, when *The Limits to Growth* triggered a huge global debate on resources, an end-of-time mood spread everywhere; people thought humanity would run out of oil and other resources in only a few decades.[11] The climate debate turned these fears upside down. Now the worry could almost be that the stocks of fossil fuels are too large.

Figure 4.7 provides an overview of the quantities of carbon contained in deposits of crude oil, natural gas, and coal—those still underground and those already exploited. Above the horizontal line, which we could imagine as the Earth's surface, are the stocks used up since the onset of industrialization; below it are the stocks still underground. For all three resource types, what is shown isn't their energy content (which stems from their contents of carbon and hydrogen), but only their carbon content, as carbon is the element that gives rise to the greenhouse effect.

Figure 4.7 again makes the distinction between reserves and resources, which was discussed in chapter 2.[12] Just as was the case when *The Limits to Growth* was published, reserves are still sufficient for 40–60 years, because the reserves are understood as the stocks whose average exploitation costs are lower than today's market price. Resources include reserves and all known stocks, even those not yet explored in depth, which at today's prices are not yet economically recoverable. These include tar sands (which are strip-mined in Canada and transformed into crude oil through a complex process that consumes a substantial amount of energy) and methane hydrates (frozen methane found in ocean beds at least 500 meters deep and in permafrost areas).[13] Unconventional resources also include tight-sand gas, shale gas, and coal-bed methane. The multiple uses of natural gas, from domestic use to powering piston engines, open up attractive possibilities of a long-term supply from such unconventional gas deposits. Averaging the numbers cited in the literature mentioned above gives a total carbon resource stock still in the ground of about 6,500 gigatons.

Since the onset of industrialization, only about 23 percent of the carbon reserves, 6 percent of the conventional carbon resources, and 5 percent of all carbon resources have been exploited.[14] The most used-up resource, at 16 percent, is crude oil; the least used-up, at about 3 percent, is coal. Natural gas is in between, at about 6 percent.

How much stays in the air?

Although so little carbon has been consumed so far, the carbon dioxide concentration of the atmosphere has already shot up to a level that has

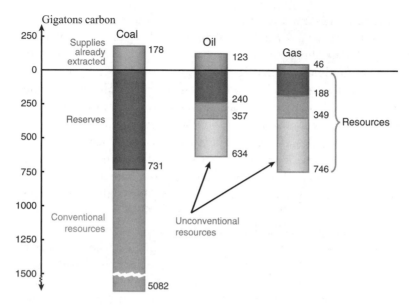

Figure 4.7
How much carbon is still underground, and how much has already been extracted? The four sources cited below refer to energy content instead of carbon volume. To construct the figure, the mean values of the fuel-specific data given in these sources were converted into the underlying carbon volumes. To this end, the molecular weight of carbon and carbon dioxide as well as resource-specific emissions factors were used for the energy equivalents, as published by the Deutsche Emissionshandelsstelle (German Emissions Trading Authority). The aggregate stock of carbon resources *in situ*, averaged over the cited sources, is 6,462 gigatons. Sources: Bundesamt für Geowissenschaften und Rohstoffe, *Reserven, Ressourcen und Verfügbarkeit von Energierohstoffen 2006* (http://www.bgr-bund.de); B. Metz, O. Davidson, R. Swart, and J. Pan, eds., *Climate Change 2001: Mitigation. Contribution of Working Group III to the Third Assessment Report of the Intergovernmental Panel on Climate Change* (Cambridge University Press, 2001), p. 236; N. Nakicenovic, A. Gruebler, and A. McDonald, eds., *Global Energy Perspectives* (Cambridge University Press, 1998); J. Goldemberg, ed., *World Energy Assessment: Energy and the Challenge of Sustainability* (United Nations, 2000), p. 149; Deutsche Emissionshandelsstelle, *Emissionsfaktoren* (http://www.dehst.de).

not been seen in 800,000 years. (See figure 1.4.) What will happen to the climate if all the rest of it is burned? That question deserves careful analysis.

Fortunately, not all the CO_2 ends up in the atmosphere. It is indeed released into the atmosphere, but it doesn't remain there. As was explained in chapter 1, a significant portion of it is absorbed by the oceans and by biomass, thereby becoming integrated into nature's cycle. The exchange between the oceans, biomass, and the atmosphere happens very rapidly. Through photosynthesis and plant decay, the carbon in vegetation is rotated, on average, every 11 years. The carbon in the upper layers of the oceans cycles within days or months, depending on the temperature, the wind speed, and the depth of the layers in question. CO_2 binds with water to form carbonic acid, falls into the ocean in raindrops, then fizzes out again, in part as a result of wave action. This cycle from the air through plants and oceans and back to the air causes the carbon stored in the atmosphere to cycle every 3–4 years on average.[15]

This quick cycle, unfortunately, doesn't mean only that CO_2 resulting from burning fossil fuels disappears from the atmosphere swiftly; it also means that it returns to the atmosphere rapidly. Only a portion of it can be absorbed in a more permanent fashion by oceans and biomass. A substantial part remains in the atmosphere for a very long time, as the exchange with ocean water takes place only up to depths of 200 meters (650 feet) or so. The saturation of water, air, and biomass with CO_2 is in an equilibrium that results in all of these three storage locations' being filled in certain proportions by the CO_2 that mankind adds by extracting fossil fuels. Only over thousands of years does the surface water mix with deeper water, increasing the portion of carbon stored in the oceans.

Figure 4.8 provides a schematic view of how the addition of new carbon affects the cycle, and of the magnitudes involved. Hydrocarbons (oil, gas, and coal) are extracted and burned. The water resulting from hydrogen combustion soon falls as rain and is of no interest in this analysis. The carbon collects partly in quick-growing vegetation and the oceans. After 100 years, 45 percent of the emitted carbon is still in the atmosphere. After that, this proportion decreases somewhat, but not by much, not least because the storage capacity of the oceans and biomass decreases as the Earth becomes warmer. After 300 years, a fairly stable final state is reached in which 25 percent of the original CO_2 remains in

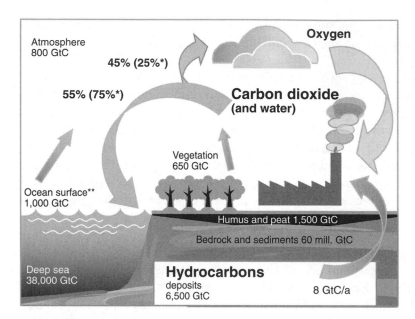

Figure 4.8
Where carbon dioxide remains. *After 300 years. **To a depth of 200 meters (660 feet). GtC = gigatons of carbon. Sources: F. S. Chapin III, P. A. Matson, and H. A. Mooney, *Principles of Terrestrial Ecosystem Ecology* (Springer, 2002), p. 335 ff.; D. Archer, "Fate of fossil fuel CO_2 in geologic time," *Journal of Geophysical Research* 110 (2005): 5–11; D. Archer and V. Brovkin, "Millennial atmospheric lifetime of anthropogenic CO_2," mimeo, 2006; G. Hoos, R. Voss, K. Hasselmann, E. Meier-Reimer, and F. Joos, "A nonlinear impulse response model of the coupled carbon cycle-climate system (NICCS)," *Climate Dynamics* 18 (2001): 189–202.

the atmosphere permanently and 75 percent stays locked in the oceans and in biomass.[16]

Further reductions occur, very slowly, because the surface water mixes with the deeper water, but the latter gets warmer in the process until it can absorb almost no additional CO_2. Despite carbon's rapid cycle between oceans, biomass, and atmosphere, it takes, on average, 30,000 years for an additional molecule of CO_2 emitted today to be removed permanently.[17]

Popular scientific optimists overlook these relationships occasionally. Man-made influence on the climate, for instance, is played down by pointing out that only 3 percent of the yearly CO_2 emissions can be attributed to human action, and that 97 percent has natural origins, including ocean water, respiration of living beings, and decay of organic matter.[18] But the fact that a growing proportion of this 97 percent also has an anthropogenic origin (that is, has been introduced into the oceans and biomass by mankind) is overlooked. Focusing on the flow of the carbon dioxide added yearly only obscures the view. What is decisive is how the carbon stocks are allocated among the various storage media and how they increase as we extract fossil fuels.

Figure 4.8 shows how much carbon is contained in the various storage media at present. Most of it—at least 60 million gigatons—is bound permanently in rocks and sediments and plays no role in the greenhouse effect. Humus and peat contain 1,500 gigatons. The dangers lurking there were noted in chapter 1 in relation to the Siberian permafrost and in chapter 3 in relation to the Indonesian palm plantations on drained bog soils. Deposits of fossil fuels, as figure 4.8 shows, still contain 6,500 gigatons. A further 800 gigatons are in the atmosphere, 650 gigatons in biomass, and 1,000 gigatons as a component of carbonic acid in the upper ocean layers down to a depth of about 200 meters (650 feet). A large amount of carbon—38,000 gigatons—is trapped in deeper waters. If we could store carbon in the deep ocean, at present temperatures it would be possible to remove from the atmosphere about eight-ninths of the CO_2 emitted. As was pointed out in chapter 2, sequestering carbon dioxide in the sea bed is discussed by scientists as a possibility. However, doing so is rightly prohibited by the UN Convention on the Law of the Sea because of possible dire consequences to marine flora and fauna.

The real question, then, is this: What consequences can we expect for the climate if the fossil carbon stocks are extracted and reintroduced into the natural cycle? Table 4.2 attempts to provide an answer. It combines the forecasts of the Stern Review (including its assertions about the level of CO_2 in the atmosphere and the resulting temperatures) with the data shown in figures 2.6 and 4.8 regarding the existing carbon stocks. The simple calculation used here is based on the formula that one part per million (ppm) translates into about 2.13 gigatons of carbon in the atmosphere.[19]

Table 4.2 includes conventional resources, unconventional resources, and reserves. The CO_2 emissions resulting from changes in land use explain the present number for carbon in the atmosphere (800 gigatons) and the corresponding CO_2 concentration (380 ppm). Further changes in land use aren't taken into account by the forecasts. Here "short-term" means within 100 years and "long-term" means after the year 2300.

Table 4.2
What extraction of fossil fuels means for our climate. Reserves and resources were estimated using the arithmetical mean of the following sources: Bundesamt für Geowissenschaften und Rohstoffe, *Reserven, Ressourcen und Verfügbarkeit von Energierohstoffen 2006* (http://www.bgr-bund.de); B. Metz et al., eds., *Climate Change 2001: Mitigation* (Cambridge University Press, 2001), p. 236; N. Nakicenovic et al., eds., *Global Energy Perspectives* (Cambridge University Press, 1998); J. Goldemberg, ed., *World Energy Assessment: Energy and the Challenge of Sustainability* (UN, 2000), p. 149. For temperature projections associated with each CO_2 concentration level, see S. Solomon et al., eds., *Climate Change 2007: The Physical Science Basis* (Cambridge University Press, 2007), pp. 825 and 798f. Data on permanence of anthropogenic CO_2 are from D. Archer, "Fate of fossil fuel CO_2 in geologic time," *Journal of Geophysical Research* 110 (2005): 5–11; D. Archer and V. Brovkin, "Millennial atmospheric lifetime of anthropogenic CO_2," mimeo, 2006; G. Hoos et al., "A nonlinear impulse response model of the coupled carbon cycle-climate system (NICCS)," *Climate Dynamics* 18 (2001): 189–202.

	Share of extracted carbon stocks (including pre-industrial stocks)	Carbon content of atmosphere (GtC)	CO_2 concentration in atmosphere (ppm)	Average temperature (°C)
Pre-industrial times	0	600	280	13.5
Present	5% (347 GtC)	800**	380	14.5
Mid-century* (according to Stern estimate)	18%	1,200	560	16.5 (15.5–18.0)
All reserves burned: 1,160 GtC (estimates ranging from 868 to 1579 GtC), short-term until 2100, 45% in the atmosphere	22%	1,320	620	16.9 (15.8–18.7)
2100 (according to the Stern CO_2 estimate)	41%	1,920	900	18.6 (16.9–21.1)
All resources burned: 6,500 GtC (estimates ranging from 5,060 to 8,980 GtC), short-term until 2100, 45% in the atmosphere, hypothetical	100%	3,730	1,750***	21.4*** (18.8–25.4)
All resources burned, long-term from 2400, 25% in the atmosphere	100%	2,430	1,140	19.6 (17.6–22.6)

*In the worst-case scenario this amount would be reached as early as 2035.

**Including carbon from modified land use.

***Outside the estimate range of the IPCC formula

Forty-five percent of the man-made CO_2 emissions will still be in the atmosphere in 100 years. After 300 years, 25 percent of the emissions will still be in the atmosphere, and they will remain there practically forever. Temperature changes relative to pre-industrial times were calculated using the formula in the 2007 IPCC Report of Working Group I (p. 818f. and p. 825), according to which the absolute temperature change depends on the relative increase in carbon dioxide concentration. The 2007 IPCC Report forecasts temperature changes from pre-industrial times only up to a CO_2 concentration of 1,200 ppm. The temperature forecast for the scenario in which all fossil fuels have been extracted and burned lies, therefore, outside the IPCC's estimation range, and should consequently be interpreted only as a rough approximation. It must be construed as a hypothetical "if-then" statement, not as a forecast.

Table 4.2 shows that before the onset of industrialization in the nineteenth century there were about 600 gigatons of carbon in the atmosphere, which corresponds to a concentration of 280 ppm and an average temperature on the planet's surface of 13.5°C (56.3°F). In the course of industrialization, to the present, about 350 gigatons of carbon, or 5 percent of the resources existing in pre-industrial times, have been extracted (see figure 4.7), bringing about an increment of 156 gigatons of carbon in the atmosphere. If we add the carbon released by deforestation and the draining of marshes, about 200 gigatons of carbon have been added to the atmosphere, equivalent to 100 ppm, raising the original 280 ppm to 380 ppm today.[20] The corresponding world average temperature is now 14.5°C (58.1°F). It must be kept in mind that the warming effect is delayed, and that it would continue over several decades even if we immediately ceased to release carbon dioxide into the atmosphere. The Stern Review's business-as-usual forecast foresees a rise in temperature to 16.5°C (61.7°F) by 2050. In the most unfavorable scenario, with the carbon dioxide concentration rising to 560 ppm (see figure 1.5), this level could be reached as early as the 2030s. This corresponds to 1,200 gigatons of carbon in the atmosphere, equivalent to 18 percent of the pre-industrial stocks' having been extracted. This isn't far from the numbers in the table's next row, which refer to the case in which all reserves have been extracted. The temperature would then be 16.9°C (62.4°F), almost 3.5°C (6.3°F) above the pre-industrial level. This situation could come about in the second half of this century.

The last three rows in table 4.2 are particularly interesting. The third row from the bottom shows the Stern scenario from figure 1.5, whereby doing nothing to curb emissions could mean a carbon dioxide concentration of 900 ppm by 2100. According to IPCC 2007, this would lead to a temperature of 18.6°C (65.5°F; point estimate). The corresponding amount of carbon in the atmosphere would be 1,920 gigatons, equivalent to extraction of 41 percent of the total resources. A temperature of 18.6°C would be 4.1°C (7.4°F) above today's level, and 5.1°C (9.2°F) above the pre-industrial level. This is the alarming scenario of the Stern Review. According to it, the Greenland icecap would melt and large portions of Bangladesh would be flooded. Stern considers 5.5°C (9°F) above the pre-industrial temperature the threshold beyond which humanity enters "unknown territory." Regardless of that, let us resume our thought experiment and consider what would happen if we extracted and burned the entire stock of carbon still underground before 2100. That absolutely unrealistic scenario helps to mark the boundaries of the theoretical possibilities. As can be seen in the next-to-last row in table 4.2, the amount of carbon in the atmosphere would then be 3,730 gigatons, which corresponds to 1,750 ppm. If we employ the estimating equation the IPCC uses for its own forecasts, we arrive at an associated average temperature of 21.4°C (70.5°F), with a fluctuation range from 18.8°C (65.8°F) to 25.4°C (77.7°F). On average, that would be 8°C (14.4°F) above the pre-industrial average—an utter catastrophe.

However impossible this scenario may be, it conveys an important message: that humanity cannot rely on the finiteness of the fossil-fuel resources in the Earth's crust to impose a natural barrier to a climate catastrophe. Nature unfortunately hasn't provided humanity with such an emergency brake. There is enough carbon underground to cause an intolerable climatic situation if we extract and burn it all in this century.

If fossil fuels were extracted slowly enough over the coming centuries to give nature the opportunity to reabsorb 75 percent of the carbon instead of only 55 percent, as shown in the last row of table 4.2, the scenario with wholly depleted resource stocks would just avoid triggering a climatic catastrophe. It would bring the amount of carbon in the atmosphere to a stable 2,400 gigatons, equivalent to about 1,100 ppm. The fact that in this case the average temperature would be "only" one degree above Stern's business-as-usual scenario for the year 2100 may mean that humanity wouldn't venture far enough into Stern's unknown territory to risk its very existence. However, as was discussed in chapter 1, that isn't at all certain, as even temperature increases of that order could trigger destabilizing secondary effects that would make the IPCC formula inapplicable. Even if the ultimate catastrophe didn't occur, the damage would be enormous. More numerous and more powerful hurricanes, a rise in the level of the oceans, and desertification of parts of the planet would impose heavy costs on humanity and force large fractions of mankind to migrate to other regions of the world, which would probably not be a peaceful process. Living on our planet would become hot, uncomfortable, and dangerous.

At the mercy of the sheikhs

Humanity has an energy problem and a climate problem. There are huge carbon deposits underground, amounting to about 6,500 gigatons, and we don't know whether we should use them. On the one hand, they are natural resources without which the wheels of industry would barely turn; on the other hand, they are lurking poisons that would make the climate unbearable in vast portions of the Earth. To find the right exploitation path—that is, the quantity and the pace that represent a compromise between these two considerations—is probably humanity's greatest task for the next few centuries.

Politicians in industrialized nations would like to set a timeline they consider appropriate for the exploitation of these resources, and perhaps they will be able to do that if a truly global Kyoto system is put in place. At the moment, however, the decision isn't theirs to make. It is up to the owners of the resources—the oil sheikhs, the Russian potentates, the large American coal corporations, and the Western oil multinationals (Shell, Exxon, BP, et al.). They decide when and how much they will bring to market, how quickly the wheels of industry can turn, and how warm our planet will get.

One cannot assume that the resource lords have the best of intentions, but neither can one assume that their intentions are malign. They just want to do well and become a bit richer and more powerful than they already are. If we want to judge whether the fate of mankind is in good hands with them, and eventually to devise policy measures that will influence their actions, we should first understand what drives them—not an easy task.

The laws that pertain to the supply of fossil fuels are entirely different from those that pertain to normal, renewable goods, because the fuels come from finite deposits and cannot be manufactured. The long-term supply from nature is as constant as in our thought experiment above, but how much the respective owners will offer in the market, and when, is up to them.

Eventually the supply will begin to dwindle, with prices trending ever upward. All relevant forecasts assume that the rate at which crude oil is extracted will peak within the next 20 years. The rest of the fossil fuels, which are available in greater abundance, will follow a similar pattern later. Figure 4.9 shows, by way of example, a forecast for crude oil extraction in comparison to the extraction rates to date. According to it, we are very close to the peak, and extraction volumes are likely to decrease in the years to come. Whether this is really so is difficult to say with certainty, as what will happen in the future depends on how the resource owners expect the consumer nations' economic development to

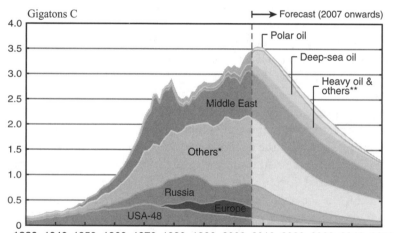

* Venezuela, Libya, Nigeria, Kazakhstan, etc.
** From oil sands and oil shale, mainly Canada and Venezuela

Figure 4.9
Historical and future oil extraction pathways. Source: C. Campbell, *Oil & Gas Depletion*, August 2008. Data from the source were converted from barrels into tons. One gigabarrel of crude oil is equivalent to 0.1247 gigatons of carbon. "USA-48" comprises the US states south of the 48° latitude. European crude oil comes mostly from the North Sea and is extracted mainly by Norway and Great Britain.

progress and on surprises that may spring up regarding reserve volumes. The owners must ask themselves, at any time, how to proceed with their remaining stocks, and should have an extraction strategy that extends far into the future.

What drives the resource owners?

The owners of the carbon deposits that fuel our world must strike a balance between their current and future extraction volumes. They could, for instance, start with high extraction volumes today, which would mean a quick decrease in later years. In that case, the prices they fetch would be low at the beginning but much higher later, and both the path of the extraction flow and the price path would be steep. After the peak, the extraction path would fall steeply, and the price path would rise steeply forever. Or they could start out with a comparatively low extraction volume and try to stretch their extraction activities over a long period. In that case, both paths would be flatter. Prices would be higher at the beginning and lower later than they would have been otherwise (though still higher than they are today).

The temporal distribution of resource extractions is a problem of portfolio optimization. It is similar to the problem routinely faced by investment bankers. In fact, the resource-extracting countries hire the best investment bankers they can find on Wall Street, and they hold regular conferences at which the most sophisticated mathematical strategies for maximizing their wealth returns are discussed. Essentially, the resource owners have to decide between leaving their wealth underground as carbon and exploiting their deposits and investing the proceeds in the capital markets. Because fossil fuels will become increasingly scarce as they are extracted, the value of unexploited deposits will increase steadily. The deposits, thus, yield returns in the form of capital gains. However, there is also the lure of capital-market returns, which can be realized only if the resources are extracted and sold for cash that can then be invested. With the help of their investment bankers, the owners divide their wealth between above-ground resources and underground resources so that the expected overall return they earn on their wealth is maximized. They will leave extraction for the future if the expected appreciation of their fossil-fuel deposits exceeds the rate of interest. Conversely, they will extract more now if they expect the appreciation of their deposits to be less than the interest they would get in the financial markets.

If all resource owners follow this simple decision principle, the price of the fossil fuels extracted will evolve, if the expectations are right, such that the rate of appreciation of stocks left underground equals the rate of interest that could be earned by investing the proceeds of the fossil fuels sold. Should this equality be upset temporarily, there are strong market forces that tend to reinstall it. If, for example, the capital market offers the more attractive conditions, most owners will decide to extract more today in order to invest the proceeds in financial assets. The increased flow of extraction will depress the current energy price and the value of fossil-fuel deposits, but the increased future scarcity resulting from this activity will increase future prices and hence increase the expected rate of capital gains on the deposits. The opposite will occur if the expected rate of capital gains on such deposits exceeds the market rate of interest. In that case, the resource owners will decide to postpone extraction to a future date. But if they do that, they will reduce today's supply and thereby push up current energy prices and the current value of the fossil-fuel deposits, and will bring future energy prices below the levels that would have prevailed otherwise and hence reduce the rate of appreciation of said deposits. This way, the markets generate a pathway in which the expected appreciation returns from resources left unexploited tend to equal the rate of interest on financial investments.

In the simplest theoretical case, where extraction costs are zero, the price of a unit of extracted resource equals the price of a unit still in the deposit. The portfolio rule then implies that the price of the extracted resource also rises at a rate that equals the capital market's interest rate. If, for example, the annual interest rate is 8 percent, the annual expected rate of capital gains for the resources left *in situ* and the price of the extracted resource must also be around 8 percent. This implication is known as the Hotelling Rule. It was named after the American economist Harold Hotelling, who in 1931 described the behavior of suppliers of exhaustible natural resources.[21]

One can also expand the Hotelling Rule to include extraction costs. Since fossil-fuel deposits aren't all equally accessible, unit extraction costs differ from site to site. These extraction costs include the investment in oil-well derricks, transport costs, exploration costs for detailed prospecting of the fields, and much more. Usually, the producers first exploit the deposits that have the lowest unit costs, then proceed to the stocks that have higher unit costs, because they can invest whatever they temporarily save in extraction costs in the capital market. Only after the stocks with lower extraction costs have been depleted do they invest in the detailed exploration and preparation of stocks with higher extraction costs. Thus, the smaller the remaining stocks in the underground deposits, the higher the unit costs of current extraction. And because extraction means that the remaining stocks diminish gradually, the unit costs will increase over time.

The dependence of unit costs on the unexploited stocks modifies the Hotelling Rule. Though it is still true that the expected rate of capital gains on the resource deposits equals the market rate of interest, the price of the extracted resource follows a slightly different rule. Its increase relative to the profit per unit extracted will now tend to equal the capital-market interest rate.[22] But the subtleties don't change the nature of the portfolio problem for owners of natural resources and shouldn't bother readers who are interested in the general principles of the economics of resource extraction.

The Hotelling Rule is an abstract, fundamental rule describing the behavior of the owners of exhaustible natural resources, particularly the owners of fossil carbon resources. But because it is embedded in a complex reality of changing conditions and complementary motivations, it is easily misinterpreted. For that reason, three qualifying remarks are in order.

First, the Hotelling Rule doesn't necessarily imply that the extracted volumes are reduced steadily in order to push prices up. That would be the case only if the demand curve didn't shift. Because general economic growth and population change shift the demand curve outward, the rule is compatible even with temporarily rising extraction flows. If, for example, as now is the case, some emerging nations (e.g., China and Brazil) grow very rapidly, resource owners tailor their supply to meet the swiftly growing demand in such a way as to achieve a growing *relative* scarcity that brings about an expected rate of price increase for the *in situ* stocks that matches the capital-market rate of interest. It is clear that they have to reduce the extraction flows at some point, because their underground deposits will eventually become depleted, but that could be done at a later date—perhaps when the emerging countries' rate of growth lessens again as they reach the levels of developed countries. Thus, the Hotelling Rule is perfectly compatible with the peak-oil scenario as depicted in figure 4.9. It is easier to understand toward the right of the peak, where extraction flows are falling fast enough to generate the necessary rate of price increase, but it also fits the upward-sloping path of the extraction curve if extraction grows slower than the world's general economic activity.

Second, the Hotelling Rule refers to expected price changes and isn't directly applicable to the actual changes. The actual changes are realized only when the resource owners correctly anticipate the evolution of the market and there are no surprises in the evolution of demand or in the information about the existing reserves. In reality, new information arrives continuously and leads to a permanent refinement and adjustment of the plans, resulting in very irregular price jumps that don't appear to coincide with the Hotelling Rule. The contradiction is only apparent, however, since the rule applies to the goals of the producers in the resource market and isn't to be construed as a deterministic forecast rule. A ship in stormy seas heading for port will be repeatedly thrown off course, but that in itself is no reason to blame the captain for lacking orientation.

Third, the Hotelling Rule doesn't apply only to competition; it also holds (in a modified version) for a monopoly. The establishment of OPEC, in 1960, was an attempt to create a monopoly. Only a fraction of the oil-producing states joined. Today the OPEC countries account for about 40 percent of the world's oil production. Even with a monopoly, the resource owners will be undecided as to whether to hold on to their reserves or to exploit them only if the rate of appreciation of their unexploited resource deposits is comparable to the rate of return they could earn in the capital market. Because, as Joseph Stiglitz showed in 1976, market power is limited in the case of exhaustible resources,[23] no fundamental modifications are needed when applying this insight to the price evolution of the oil already extracted. Market power presupposes that one can push the price up by withdrawing some supply. Though a monopolist has this power in the case of exhaustible resources, he will have to increase supply in the future if he doesn't want to get stuck in the long run with part of his resources unsold. From this point of view, OPEC is a toothless tiger. Time and again it has tried to force prices up, and since World War II it has provoked two oil shocks by doing so. But in the end it has always failed, subsequently having to accept significant price concessions. In the case of the market for exhaustible resources, focusing attention explicitly on market power leads to no particular insights into the long-term evolution of prices. Thus, we may as well focus on the case of competitive markets.

Greed and sustainability

Can mankind feel comfortable with the goals of the owners of fossil-carbon resources? Are those resource owners, who can make practically the entire Western world dance to their tune, doing the right thing, or are they extracting their resources much too quickly? That they are trying to do the right thing for themselves is clear. But can the market behavior described stand up to a welfare assessment from the society's point of view?

Critics of capitalism may feel tempted to point out that resource owners are driven by greed and for that reason are fundamentally unable to extract the resources in a conservative manner. But this position is untenable simply because the pursuit of individual self-interest is the general secret behind the success of a market economy, something that has been amply demonstrated by the historical superiority of this economic system and by the theoretical work of Kenneth Arrow and Gérard Debreu.[24] The market economy manages scarce resources fundamentally better than other economic systems. Although communism was incapable of providing the population with more than a basic level of goods and services, it exploited natural resources ruthlessly and had no qualms about polluting the environment around its industrial combines to an appalling degree. Today, after the triumph of capitalism over communism, the economies of the formerly communist countries are more efficient than ever before, yet their consumption of fossil fuels is markedly lower than during their communist past. The dramatic decline in Russia's emissions (see figure 2.1) is a truly poignant indictment of the communist system.

As the considerations of the portfolio problem in the previous section showed, the pursuit of profit can indeed lead to low-impact, sustainable resource extraction. Ruthless exploitation cannot be triggered by mere profit seeking, because such extraction practices would make resources scarcer and therefore more expensive in the future, which would promise higher returns from keeping the resources underground. A resource-conservation strategy would bring higher returns to capital than rapid extraction coupled with an investment of the proceeds in the capital market. Thus, the profit motive in itself clearly doesn't contradict sustainable resource management.[25]

A further reason for skepticism that might occur to critics relates to the altruism that resource owners feel toward their heirs. If they leave carbon underground, they do it with a view to leaving it to their offspring. But their heirs are, of course, only a tiny portion of mankind. Can we assume that the private altruism of the sheikhs toward their heirs is sufficient to safeguard the interests of future generations as a whole? The value of the oil reserves of the Gulf Cooperation Council (GCC), which comprises Kuwait, Bahrain, Saudi Arabia, Qatar, the United Arab Emirates, and Oman, and which accounts for about 40 percent of the world's oil reserves, is estimated at $65 trillion. The value of all the oil reserves in the world is estimated to be equal to the value of all the houses and all the factories on the planet. The resource owners administer truly gigantic wealth. What if the sheikhs and the other resource owners pay less heed to the future than the rest of mankind considers appropriate?

At first sight, a fundamental problem appears to manifest itself—one that offers grounds for concern. When the situation is examined more closely, however, it is clear that this isn't the case. If the owners of the oil fields were to forsake their children and grandchildren in order to grant themselves a first-class standard of living now, they wouldn't have to exploit their reserves more quickly at all. They could simply sell their fields. After all, in many cases the property is already in the hands of corporations in which they own shares. Selling, thus, would be a simple thing. After the shares had been sold, the new owners would set the extraction policy, and the portfolio optimization would be carried out as described above, but nothing fundamental would change, as the question of altruism doesn't figure at all in the comparison between the rate of appreciation of the resources *in situ* and the capital market's interest rate.[26]

Leaving that aside, we can, of course, bemoan the unequal distribution of wealth in the world, and wish that we, instead of the oil sheikhs, owned the resources. That is, however, an entirely different matter. The question of distribution has nothing to do with the question of whether the sheikhs and other potentates will exploit their resources at the right pace.

Nirvana ethics

Altruism toward one's descendants plays no role in the rate of resource exploitation chosen by market forces when the rate of interest is given. However, it could affect that rate. The more altruistic people are in general, the more they save, and the lower is the market rate of interest. The focus here is not on resource owners, though, but on all savers of the world, of whom the resource owners are only a fraction. A lower rate of interest means that it is more attractive for resource owners to leave more fuel underground and hope for capital gains. This decision, however, increases the current price of the resource and lowers the future price relative to what it otherwise would have been, which implies a lower rate of price increase—in equilibrium, one that equals the lower rate of interest.

Some assert that we are bequeathing too little to our descendants, and hence save too little, because we are too selfish and discount the interests of future generations too strongly. One can reason along these lines from an ethical-philosophical point of view.[27] The Stern Review, which also takes this philosophical position, argues vehemently that present generations are not according their descendants the consideration they deserve.[28] With the interest mechanism described, this could indeed be the case. If the argument is right, the rate of interest is too high in the world capital market, and hence resources are extracted too quickly.

However, this line of reasoning doesn't lead very far when it comes to policy measures to be taken. If we bequeath too little to later generations because we aren't sufficiently altruistic toward our descendants, who should we entrust with the decision as to the correct measure? The government, perhaps? That wouldn't help. Its leaders have, after all, been elected by the current generation, and will therefore be no less disinclined to accord later generations more importance than any individual deciding on his or her own consumption. We would have to appoint as guardians of the interests of future generations those philosophers who accuse people of too little altruism toward future generations. But how would that work? Nothing short of granting them dictatorial powers that they could exert against the democratically sanctioned will of governments and parliaments would have a chance of achieving anything. Such an argument must have arisen from the realm of nirvana ethics, as it bears no relation whatsoever to the political ways of a democracy.

The representatives of future generations are, in any case, the people currently living (that is, us), and whatever consideration we accord future generations has to be taken as such, because there is no possibility of making future generations take part in such decisions. Since they aren't here yet, there is no other way than for us to decide in their stead, whether these decisions are made by the market or by politics.

Incidentally, the fear that we would not accord future generations sufficient consideration in our moral deliberations can be called into question on the ground that future generations don't come from Mars but rather are our own children and their children. All people who will be affected by the decisions we make today are represented here and now by their ancestors. Evidence of insufficient affection for descendants or of their inadequate representation in today's decision processes is hard to detect.

This should not be taken to mean that we can be content with the intertemporal extraction speed for fossil carbon resources, whose waste causes global warming. It just means that the ethical argument doesn't lead very far.

Wrong expectations

Similar reservations are appropriate with regard to the argument that government intervention is necessary to smooth the extraction and price paths chosen by markets, which often seem rather erratic. It is true that the resource owners often err in their appraisals of future demand, and that this leads to irritating ups and downs in the evolution of prices resulting in a deviation from the "true" Hotelling path that would emerge in the case of correct expectations. But such failings would legitimize a political intervention only if it could be assumed that politicians know better than resource owners how prices and demand will evolve.

Such an assumption isn't plausible. No one is better informed about market trends than the resource owners, who are betting their own money on it. The lack of information about the future that leads to erroneous expectations is indeed a problem, but that problem is probably even bigger for politicians than for participants in the private market. An alleged need for action on the part of the world community's collective organs in the area of resource exploitation can't be derived from such arguments.

Once again: In no way does that mean that there is no need for action, and that the exploitation pathway chosen by the market forces is the correct one. It would be wrong to interpret the chain of arguments presented above as an attempt to blindly appraise market forces. What really happens in the resource market doesn't depend solely on values and expectations; it depends, most of all, on the rules of the economic systems in which the resource owners operate. Here lies the snag. The problem isn't that people *want* the wrong thing; it is that they *do* the wrong thing, because they operate in economic systems that don't present them with the real costs of their decisions. The following sections will attempt to make clear, step by step, where the market failures may occur.

The social norm

To evaluate market behavior from the point of view of welfare economics, we first need a norm. Society as a whole, after all, faces a portfolio problem, just as resource owners do. We can bequeath our descendants wealth. We can build houses and factories that they can use after we are dead. We can also bequeath them the natural resources that we inherited from our ancestors, an intact environment, and a suitable climate. Put simply, we have to strike the right balance between man-made capital above the ground, natural capital below the ground, and carbon waste in the atmosphere. For the moment, let's leave aside the problem of carbon waste in the atmosphere and focus on the capital above and below ground when there is no waste in the atmosphere.

Whether we wish to bequeath our descendants capital underground or capital above ground, both require people living today to relinquish some consumption today. If we devote production factors to expanding man-made capital by building more machines, houses, and factories, these factors will not be available for the production of consumer goods. Likewise, if we leave more carbon underground (while keeping the investment in man-made capital unchanged) we will lack energy for the production of consumer goods. The conflict between our present consumption and the consumption of future generations is unavoidable. But precisely because that is so, present generations should aim at extracting the highest possible gain out of relinquishing consumption in favor of future generations, regardless of how much consumption they are ready to forgo. This is a weak social norm, but one that everyone should be able to accept.

From a social perspective, the gains from relinquishing consumption with the purpose of bequeathing man-made or natural capital to future generations consist of increasing future GDP. If we bequeath future generations more man-made capital, they will be able to generate more output with their labor, because they will be able to work with more sophisticated and costly equipment and a better infrastructure. We have profited in a similar way from the frugality of earlier generations. The fact that we can enjoy a much higher standard of living than our forefathers stems from our being able to leverage our work much more strongly because the stock of man-made capital that is available to us is greater than the corresponding stock that was available to our ancestors.

The market rate of interest is a good measure of the increase in future output made possible by the larger stock of man-made capital, as it shows what entrepreneurs can gain beyond recouping their capital investment. Indeed, it can be shown that the interest rate in a competitive economy indicates exactly how much additional output net of capital depreciation will be generated in the future by a given investment in capital. This refers not only to the annual interest rate but also to compound interest. A 3 percent annual interest rate translates in ten years into a 34 percent interest rate. If we were to forsake consumption to the tune of a billion dollars today, we would make possible a one-time $1.34 billion increase in consumption a decade from now at today's prices. If the capital were left untouched, consumption in all following decades could be $340 million higher than it would have been otherwise.

If instead of bequeathing more man-made capital we bequeath to future generations more natural capital in the form of a larger stock of fossil fuels *in situ*, that will also increase the future GDP, because more energy will be available for production. As this will mean sacrificing some output today, an economic return will result from this strategy if in the future more additional GDP can be generated from a further ton of available fossil carbon than can be produced from that ton today. Given that natural resources become scarcer over time, it can indeed be expected that that such will be the case. At present, a sizable amount of energy is wasted because it is still relatively cheap. In the future, when energy is scarce and expensive, we will think twice or three times about how and where to use it. If a bit more carbon is available then, it will make a larger contribution to GDP than is lost today. Economists describe this phenomenon by saying that the increasing scarcity of carbon results in an ever-increasing *marginal product* of carbon. The percentage at which the additional future output surpasses today's loss in output is the social rate of return of the carbon left *in situ*. If there are extraction costs, a slight modification of that statement is called for. The relevant figure for calculating the economic rate of return of the resource left *in situ* is now the growth rate of the contribution to GDP, net of extraction cost, that an extra ton of carbon could generate today or in the future.

If the economic rate of return achievable through a deferral of resource extraction lies above the capital-market interest rate, we should indeed defer the extraction from a social perspective, because that could be part of a strategy to make future generations better off without having to lower our standard of living. In the language of economics, we could generate a Pareto improvement to the benefit of future generations—an improvement that costs us nothing. The strategy is changing the composition of wealth we bequeath to future generations without changing the size of the bequest. It is true that postponing the extraction of fossil fuel reduces today's GDP, because fossil fuel is an important production factor. However, we don't have to reduce today's consumption; we could simply reduce the amount we save and bequeath in terms of man-made capital out of the reduced GDP, giving future generations less intensely capitalized factories, less infrastructure, or simpler buildings. Future generations would profit from such a strategy in the same way as someone who has bought shares in an investment fund and sees higher returns when his fund manager shifts the funds he has been entrusted with from an investment earning a low interest rate to one earning a high interest rate. That person will later enjoy a better standard of living without having to sacrifice anything in the present.

If we are selfish and don't wish to make future generations better off, we can even translate the improvement in the portfolio composition into an increase in our own consumption without hurting future generations. To do this, we simply have to reduce our bequest of real, man-made capital a bit more so that the standard of living of future generations will just stay unchanged while our own consumption rises. We give up GDP because we extract less carbon today, but we expand our consumption by reducing the part of GDP that is used for investment in man-made capital. The decrease in savings and bequest, and hence investment in man-made capital, absorbs both the decline in GDP and the increase in present consumption. Future generations just produce a given GDP with

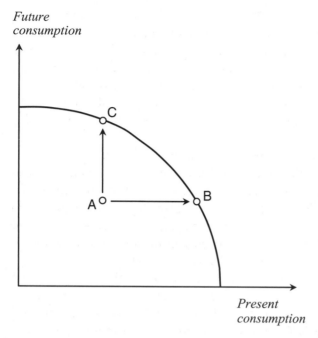

Figure 4.10
Intertemporal Pareto optimality

more carbon and less man-made capital as before the portfolio shift, and thus they can afford the same consumption and investment as before. The benefit for us would be similar to that of the investor who takes advantage of the improved investment strategy of the fund manager by withdrawing money from the fund for immediate consumption without sacrificing his future standard of living.

Figure 4.10 illustrates the two possibilities for Pareto improvement. There are many ways to transform present consumption into future consumption, depending on the portfolio mix between man-made and natural capital and the overall bequest volume present generations are willing to sacrifice. Each of these possibilities is represented by a point below or to the left of the curve. The curve itself is the efficiency frontier of the technically feasible strategies. Obviously, any rational society should want to be on the curve. If it were below the curve, it could, by changing its portfolio strategy, increase consumption in the future, holding consumption in the present constant, or vice versa. It could even increase consumption in both periods.

Suppose the rate of return from postponing fossil-fuel extraction is higher than the market rate of interest. In that case, mankind finds itself at an inefficient point, such as A in figure 4.10. If present generations bequeath more natural capital and less man-made capital, keeping their own consumption constant, the economy will move up vertically in the diagram toward a point such as C. However, by reducing the bequest of man-made capital a bit more so as to keep future consumption unchanged, present generations could also move from A to B. Both of these movements would be Pareto improvements.

The second possibility in particular shows that a resource-conservation strategy will not necessarily imply any sacrifice for the present generations and doesn't require intertemporal altruism. If indeed the rate of return from postponing extraction is higher than the market rate of interest, as was assumed, even selfish decision makers acting today should be "green" conservationists.

Unfortunately, we can't go beyond the efficiency frontier. The more we want to exploit the difference in the rates of return by moving toward the efficiency frontier, the smaller it will get, and when we are on the efficiency frontier it is zero. The market rate of interest is higher the less mankind invests in man-made capital, as less profitable investment projects are given up first, and the rate of return from postponing the extraction of fossil fuels is lower the more the extraction is postponed. For one thing, the wasting of carbon today is reduced if less carbon is available. For another, the increase in the carbon supply in the future implies that carbon will be dealt with less carefully then, so that additional tons of carbon will generate lower and lower increments in GPD. An optimum exploitation path from an economic perspective is one in which the economic rate of return through deferred exploitation will always equal the capital-market interest rate, because then a small shift in society's portfolio composition is unable to make one generation better off without forcing another one to sacrifice consumption. In figure 4.10, this would be a point on the efficiency line such as B or C. In economics, this condition is known as the Solow-Stiglitz Efficiency Condition, after Robert Solow and Joseph Stiglitz.[29]

The condition for Pareto optimality is very similar to the one observed by private resource owners in their extraction plans. In fact, in a well-functioning competitive economy without externalities, and particularly without the global-warming externality, the conditions are identical. In a market economy, the contribution made to GDP by the last ton of carbon extracted in a given period equals the sales price of that ton. Furthermore, the extraction costs have the same meaning in the private and the social calculation. In each period, therefore, the market value of a unit of carbon *in situ* is a correct measure of the increase of GDP, net of extraction cost, potentially available for consumption that would be generated by extracting one further ton of carbon. From this it follows that the time path for resource consumption is Pareto efficient when the rate of appreciation of the resource *in situ* corresponds to the market rate of interest, which is exactly the Hotelling Rule that private resource owners follow in competitive markets.

Why extracting more slowly makes the cake bigger

In global warming, extracted carbon accumulates in the atmosphere and causes damage around the world that doesn't figure in the economic calculations of either suppliers or consumers of fossil carbon and therefore remains unacknowledged by them. The market failure resulting from these external effects is, in the opinion of the Stern Commission, the greatest market failure ever. That assertion is very difficult to disagree with.[30]

At first glance, the appropriate response seems clear: Insofar as carbon in the atmosphere causes higher temperatures, which are damaging, we shouldn't permit as much carbon to reach the atmosphere as we have permitted. For many political purposes, this insight is sufficient.

But this is too imprecise to provide a scientific foundation for policy formulation and correct the above market failure. It is not clear, for instance, whether we should merely dampen the zeal of the market in order to slow down extraction, or whether a portion of the fossil-fuel resources should be removed forever from any economic use at all. Must we then seal some deposits off permanently, barring access even to their owners, or will it suffice to find ways to induce owners to extract carbon from their stocks more slowly?

Some may find this distinction academic. But it isn't so. Both from the perspective of the owners of fossil-fuel stocks and from that of the possible policy options for correcting the market failure, this distinction is fundamental. If deposits were to be sealed off, they would have to be expropriated or bought in order to place them, akin to nature reservation or preservation of cultural heritage, under the protection of the UN. If, however, extraction should proceed more slowly, adequate financial incentives or clever management through global permit trading may suffice.

From the perspective of the portfolio problem described above, it is possible to justify the postponement of extraction if one takes into account that the damages accruing to society as a result of higher temperatures would also be deferred to the future.[31]

Each decade in which the extraction of a given amount of fossil fuel is postponed is another decade without this amount causing damages and repair costs. It means one less decade of extra droughts, floods, storm damages, and it means lower costs for cooling our homes, offices, and cars.

The economic gain from deferring some extraction to the future results not only from an increase in future GDP beyond what it costs in terms of GDP reduction today (because increased resource scarcity makes the marginal product of the resource rise over time), but also from the fact that a smaller portion of future GDP will have to be devoted to repairing the damage and hence a larger portion will be available for private consumption. Thus, society's portfolio optimum in the Pareto sense requires that the sum of these two advantages be equal to the market rate of interest.

Markets, however, don't meet this requirement, as they follow the Hotelling Rule and hence just equate the first item of this sum to the rate of interest, as was shown above. Thus, in an intertemporal market equilibrium, when the private investment bankers have optimized the wealth portfolios of resource owners, the social advantage of bequeathing fossil carbon left underground exceeds the social advantage of bequeathing man-made capital by the reduced damage from global warming. To be more precise, the percentage social rate of return from postponing the extraction of fossil fuels exceeds the market rate of interest by the climate damage that could be avoided by keeping an additional value unit of the fossil carbon in its underground deposit.[32]

This means that society is indeed at a point similar to A in figure 4.10. By extracting more slowly and bequeathing more natural capital to future generations at the expense of man-made capital, it would indeed be possible to make future generations better off without hurting the present generation. This would be a move to point C.

However, society could also move to point B. It could do that by bequeathing future generations even less man-made capital and increasing its own consumption in a measure that just keeps the GDP of future generations, net of the damage from global warming, intact. Because future generations have the double advantage of enjoying more carbon underground and less carbon in the atmosphere (and hence less damage), we can, without hurting them, curtail our bequest of man-made capital and increase our own consumption. Even though we give up GDP because we extract less carbon, we can increase our own consumption, because the reduction in capital investment exceeds the decrease in GDP.

It would, of course, also be possible to move to a point between B and C on the efficiency frontier, so as to make both future and present generations better off—a win-win situation for "green" policies that aim at saving carbon and slowing extraction relative to what the market economy brings about. The cake simply gets bigger.

A useful interpretation of the market failure that prevents the cake from being maximized can be drawn from seeing the extracted carbon as a complement to man-made capital. After all, carbon has been the main driver of industrialization in most countries. Suppose we want to build up more man-made capital without relinquishing today's consumption. The way to do this is by extracting and burning more carbon to increase GDP, and to use the extra GDP for additional capital bequests to future generations. The extraction of natural capital from underground, the buildup of man-made capital above ground, and the accumulation of carbon dioxide waste in the atmosphere are thus closely related processes. Seen in this light, the damages caused by carbon dioxide push the true economic rate of return on man-made capital, to which society should adjust the appreciation rate of the resources *in situ*, below the market rate of interest. Pareto improvements can therefore be generated by making the carbon-extraction path and the price path flatter than the paths chosen by market forces.

Figure 4.11 illustrates this interpretation in a schematic fashion for the carbon price path. The middle path shows the competitive market equilibrium, satisfying the Hotelling Rule. In such an equilibrium, the extraction path will be chosen so that the increasing relative scarcity of the extracted carbon allows the price to increase at such a rate that the resulting appreciation of the resource *in situ* equals the market interest rate. (Neglect the upper path for the time being. It results from insecure property rights, as will be explained in the last section of this chapter.)

The lower path depicts the Pareto optimum. It is flatter, because the appreciation rate of the resources *in situ* is equal to the market interest rate minus the climate damage expressed as a percentage of the capital

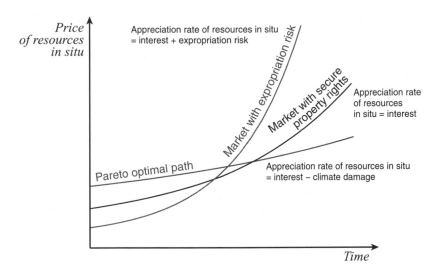

Figure 4.11
Market behavior relative to the social optimum—too cheap today, too expensive tomorrow.

investment. The Pareto-optimal path is flatter than the market path and intersects it at one point. It must start out at higher prices; otherwise, over time, more carbon would be in demand than on the market path, which would not be feasible, as at some stage there would be no more carbon left.

Relative to the Pareto-optimal path, the price on the market path is too low at the beginning and too high later. Thus, the stocks are depleted too rapidly. This provides the fundamental rationale for a policy aimed at slowing the extraction of fossil carbon, thus slowing global warming at the same time.

Why carbon deposits should not be sealed off

That carbon extraction should proceed more slowly doesn't mean that we should forgo extraction entirely; it means that extraction should be deferred to a later date. Whether part of the resources should never be extracted, on account of the climate problem, is a question that has no clear-cut answer and must be left to future generations.

To abstain from exploiting part of the resources forever would entail sealing them off and treating them as if they didn't exist. It is easy to demand something like that today, because there is still enough carbon underground. But how will our descendants in the distant future, perhaps many thousands of years from now, view this? To know that our descendants will have no interest in the sealed-off stocks because of the climate problem, we would have to know what conditions will prevail in the world at that time, how willing our descendants will be to pay, how high the extraction costs will be, and what additional damages might result from the extraction at that time. From a social perspective, it makes sense not to exploit the last oil field until some future date, when consumers' willingness to pay will be high enough to cover not only the extraction costs but also the economic costs of an even stronger greenhouse effect. From a private business perspective, on the other hand, a field will already be recoverable for the owner if the price consumers are willing to pay exceeds the extraction costs, since the owner will not have to pay for the global-warming externality. Seen in this light, it may well be the case that there are fields that will be exploited someday as a result of market forces, but which from an economic perspective should never be touched. Today, however, we know nothing about this. It is also possible that when the deposits near depletion consumers will be willing to pay prices that would exceed the extraction costs for the last oil field by more than the additional damages caused by even more global warming. In this case, the market forces would point to extraction of the entire stock, and this would be the correct decision from a social economic perspective as well. Both cases are possible, but because they both refer to a distant time with unknown economic conditions, unknown marginal damages, and unknown extraction costs, we simply can't decide today which of them is more relevant. For this reason, policies that go beyond slowing extraction simply aren't justified today. The case for sealing off fields or deposits can't be made as rigorously from a welfare perspective as the case for slowing extraction.

One could argue that consumers' willingness to pay for energy has an upper limit set by the cost at which "green" backstop technologies, from wind farms to nuclear fusion, could deliver the energy. That would make all the resource deposits in private hands that are just profitable because their unit costs are slightly below the price of the backstop technologies uneconomical when the costs of the damage caused by the greenhouse effect are taken into account. Using this argument, one could conceivably attempt to justify the sealing off of the fields economically. This argument is, however, not convincing. Of course there are alternative technologies that dampen the price increase. The higher the prices of fossil fuels, the greater the incentives to avoid consuming such fuels and to resort to alternative sources, such as solar or nuclear energy. The entire spectrum of the consumption-dampening measures listed in table 4.1 will be acted upon as soon as the prices rise, which in turn will work toward dampening the price hike. Recall that the continuum of backstop technologies for different energy uses was the main explanation lurking behind the demand curve in figure 4.4. But the alternative technologies must be *perfect* rather than partial substitutes in order to constitute a firm upper boundary for the price of fossil carbon and thus justify sealing off the carbon deposits. Such perfect alternative technologies are not in sight.

Nuclear fission, nuclear fusion, solar power, hydro power, and wind power all result in the production of electricity. But electricity can by no means be considered a perfect substitute for fossil fuels, simply because fossil fuels are superb energy-storing devices. No other substances or technical devices that could be used in transportation are able to store as much energy per unit of weight. As was argued in chapter 3, even the lithium-ion batteries used in laptop computers don't store more than 1.7 percent of the energy of diesel fuel per unit of weight. Hydrogen is a potential candidate that can be produced from electricity; however, as was also explained in chapter 3, hydrogen must be highly compressed or cooled before it can be stored and transported, and even then it requires much more space than fossil carbon.

The only perfect substitutes for fossil fuels are biofuels, which have a very similar chemical structure and can be stored and transported with identical ease. But they will not be able to set an upper price boundary to carbon fuel, because they require enormous areas of land. Replacing the fossil fuel used by transportation alone would require all of the planet's arable land. Even if BtL (biomass-to-liquid) can be developed, substantial areas of land will be needed—much more than mankind can ever afford without risking a worldwide recurrence of the Tortilla Crisis, this time with an even greater impact. Above all, even if more and more land is used for the production of biofuels, that would not mean that there is a hard ceiling for the carbon price, as the price of agricultural products would also increase (perhaps without limit). In chapter 3 it was shown why the notion that biofuels can be produced in sufficient volume to allow humanity to desist from extracting a portion of its carbon stocks forever without engendering substantial if not outrageous food price increases is simply absurd. Thus, a perfect substitute for fossil fuel is not in sight, and in the absence of such a substitute a policy of sealing off carbon deposits cannot be recommended from a welfare perspective.

The fear of a coup

A further argument for slowing carbon extraction instead of sealing off some of the carbon deposits involves the fear the sheikhs and other resource owners have of their rivals. It is often the case that the owners of natural resources are uninterested in a conservative extraction strategy because they live in permanent fear that they, their descendants, or their clans will be ousted before extracting the resource.

It is all too possible that a rival regime will come to power through a revolution, or that the Americans will come and threaten the ruling class with Western-style democracy. Insecurity is high, with a constant fear that the clan, the children, or the grandchildren will not enjoy the proceeds of exploiting the natural resources. In view of this, the resource owners consider it advantageous to exploit the deposits as quickly as possible and stash away the proceeds in Swiss banks or elsewhere.[33] The upshot is that extraction proceeds much too quickly.

This distortion in market behavior again has nothing to do with the strength of the altruism resource owners feel toward their descendants, though at first sight it might appear to. As was argued above in the section on greed and altruism, lack of altruism may lead resource owners to squander their wealth, but it doesn't imply that the stocks will be extracted too quickly. Those who wish to squander their wealth need only sell their ownership titles in the natural resources. That way, they would have the cash that the resources are worth at their disposal to spend as they see fit, but the resources would still be underground and would then be extracted by someone else. The statement that the market chooses an extraction path that equates the expected appreciation rate of the *in situ* resources with the capital-market interest rate is entirely independent of the consumption wishes of the owners of the resource.

The risk of losing the ownership of the resources *in situ*, on the other hand, does lead to severe distortions in the extraction path. One can't avoid this risk through divestiture, as the new owner will be equally affected by it. Those who own the resource face the risk of losing ownership through a political revolt, so they will attempt to extract the resource more quickly than they would if their property rights were secure.

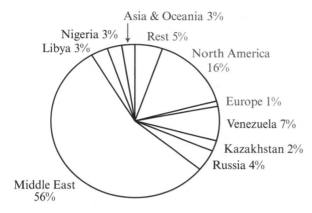

Figure 4.12
Places where oil is in the hands of "barons." Source: Energy Information Administration, *International Data 2009, Petroleum (Oil) Reserves & Resources* (http://www.eia.doe.gov).

The problem of insecure property rights doesn't apply equally to all fossil-fuel deposits. For instance, there are large coal deposits in the United States and Australia, where the risk of political upheaval and the loss of property rights is rather low. But it does apply to a portion of the natural gas deposits and, most of all, to the largest part of the oil deposits. Figure 4.12 shows where the world's oil reserves are located and who controls them. Three-fourths of the world's oil reserves are located in Venezuela, Kazakhstan, Russia, the Middle East, Libya, and Nigeria. These are mostly countries and regions with extremely insecure political situations and, consequently, insecure property rights for the resource owners.

In fact, the oil-producing countries were affected by extreme political turbulences in the last few decades that, if not directly obstructing extraction, stoked the owners' fear for their property rights. The second oil crisis coincided with the toppling of the Shah of Iran by Ayatollah Ruhollah Khomeini in 1979. From 1980 to 1988, Iran was engaged in a bitter war with Iraq. In 1990, Iraq invaded Kuwait; a year later, it was pushed back by American troops. In 1992, a civil war was raging in Algeria, and in Venezuela Hugo Chávez attempted a coup. In 1989, the Soviet Bloc began to fall apart. In 1991, in the Soviet Union, elements of the Communist Party attempted a putsch against Mikhail Gorbachev. In 2001, al-Qaeda destroyed the World Trade Center, and in 2003 the United

States invaded Iraq. In 2002 there was an attempted coup against Chávez, who by then had come to power democratically in Venezuela. At this writing, a "Jasmine Revolution" is under way in the Arab countries. All these events contributed to increasing fears of losing property rights over resources *in situ* and thus increased the willingness to extract.

The fear of expropriation brings a fundamental change to the equilibrium of the oil market, as the accelerated pace of extraction gives rise to very low prices in the present but much higher prices in the future when the oil fields are nearly depleted. For a new equilibrium to emerge in which resource owners are indifferent as to present and future extraction, the rate of price increase for the resource *in situ* will now have to be large enough not only to match the market rate of interest but also to compensate for the risk of expropriation. In mathematical terms, the rate of price increase of the resource *in situ* will be equal to the rate of interest plus the annual expropriation probability. Given that, from a social perspective, the rate of price increase should be equal to the market rate of interest minus the greenhouse damage per unit of capital, this alteration in the market equilibrium rule obviously goes in the wrong direction, exacerbating an existing distortion.

The upper path in figure 4.11 illustrates the resulting price path. It climbs even more steeply than the middle path, which represents the case with secure property rights and which is already too steep from a social perspective because markets neglect the greenhouse externality. Thus, obviously, the neglect of the greenhouse externality and the expropriation risk both work in the same direction, resulting in a too rapid extraction of fossil-fuel resources. This makes the case for "green" resource-conservation strategies even more compelling.

The insecurity of property rights could be the reason why crude oil is the fossil fuel with the shortest running time. As was shown in figure 2.6, the reserves and the conventional resources are sufficient for only 60 years, and even if the unconventional resources are added the oil deposits will be depleted in 140 years at today's extraction rates. The running times of natural gas and coal are much higher. Natural gas also occurs in places where property rights are insecure, including Siberia and Kazakhstan. This insecurity, however, isn't comparable to that affecting oil in the Middle East. The world's coal reserves, in comparison, lie in fairly safe locations. There are large deposits in the United States, China, and Australia, where no comparable insecurity prevails. The fact that 16.2 percent of the oil once available in the Earth's crust has already

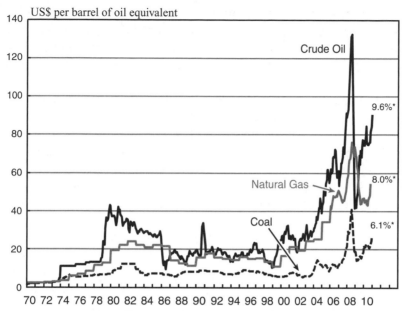

* Average annual price increase from whole year 1970 until October 2010

Figure 4.13
How the prices of oil, gas, and hard coal have evolved. This graph shows the monthly world prices for crude oil and hard coal, as well as the annual average (as of 2006 monthly) price of gas imported by Germany. The last month shown is December 2010 for the oil and coal prices, October 2010 for natural gas. Source: Hamburgisches WeltWirtschaftsInstitut, *Rohstoffpreisindex* (http:// hwwi-rohindex.org); Statistik der Kohlenwirtschaft e.V., *Entwicklung ausgewählter Energiepreise* (http://www.kohlenstatistik.de); Deutsche Bundesbank, *Statistik, Devisenkurse* (http://www.bundesbank.de).

been extracted and used, versus only 5.8 percent of natural gas and 3.4 percent of coal, is directly attributable to differences in the security of property rights. (See figure 4.7.)

Figure 4.13 illustrates the evolution of nominal oil, gas, and coal prices. Obviously these prices are extremely volatile. This could be due to expected and unexpected changes in the expropriation risk or to myriad other factors that also affect the extraction patterns. However, despite the volatility, there are interesting differences. From the average of 1970 to October 2010 the price of crude oil increased by 3,978 percent, that of natural gas by 2,141 percent, and that of coal by 987 percent. The corresponding annual average growth rates of these prices were 9.6 percent, 8.0 percent, and 6.1 percent. As was to be expected from the differing expropriation risks, oil topped the list, with gas and coal following. Interestingly enough, the average nominal interest rate for a repeated investment in ten-year government bonds over that period (to the end of 2010) was 7.2 percent.

The fundamental feature that distinguishes a market economy from other economic systems is that, through trading, it largely exploits the scope for efficiency enhancing changes in the allocation of economic resources. The precondition is, of course, that property rights are well defined and guaranteed, so that trade can take place. That is why such rights are firmly anchored in the constitutions of the European Union, the United States, and many other developed countries. In the absence of protection of property rights, trade will not exhaust every possibility to improve efficiency, since one can't trade things one doesn't own. In the case of the oil resources, this very general insight takes a concrete form that has become a major problem for the future of mankind. Oil is indeed squandered by mankind because struggles over its property rights have destroyed the conservation motive. Even if that were not the case, oil would, like the other fossil-fuel resources, be extracted and warm up the earth too quickly because markets disregard the greenhouse externality. The warnings that the Club of Rome expressed in early 1970s and the more recent warnings from the Stern Commission reinforce and complement each other in a disquieting fashion, although—or perhaps precisely because—they analyze the problem from different perspectives.

to Martin Luther, as a token of thanks for the apple tree

5

Fighting the Green Paradox

The impotence of politics

There is no doubt that mankind has to do something against climate change. At present, as was shown in chapter 1, there is more carbon dioxide in the atmosphere than at any time in the past 800,000 years, and the record for the highest average temperature is about to be broken. Before industrialization, the average temperature was 13.5°C (56.3°F); now it is 14.5°C (58.1°F), and by 2030, according to the relevant forecasts, it will break the 15.3°C (59.5°F) record of the past 800,000 years, which was reached 125,000 years ago during the Eem Interglacial warm period. Polar ice is already melting, the Greenland icecap will follow, storms will become fiercer, and deserts will expand. The resulting migrations will endanger the political stability even of countries that would otherwise have nothing against somewhat warmer weather.

We shouldn't pin our hopes on the prospect that stocks of fossil carbon yet to be exploited will prove too small to trigger a climate catastrophe should we burn them all to tap their energy. Since the onset of industrialization, barely 5 percent of the pre-industrial carbon stocks have been extracted; if the rest should be extracted in an uncontrolled manner, 18 percent will have been extracted by the middle of the century, and 41 percent by 2100. The average temperature would then be expected to reach 18.6°C (65.5°F). Further extraction over the coming centuries would, under the same assumptions, push the average temperature to nearly 20°C (68°F), and it would stay at that level for a very long time. That is about as much of an increase from today as occurred between the coldest point in the latest ice age until today (5.5°C, 9.9°F), but in fast motion. With even faster extraction, the average temperature might temporarily move above 21°C (69.8°F).

In view of these prospects, political bodies the world over are under great pressure to act. But what can they do? Other than afforestation and sequestration (which is problem-laden), all they can do is try to reduce carbon extraction. There is no alternative, because of an unavoidable fact of which many people aren't aware: There is no technical lever that would allow the burning of a certain amount of carbon with higher or lower CO_2 emissions. The carbon that we extract from the ground in order to gain energy from it through combustion is the same carbon that subsequently reaches the atmosphere in an oxidized state. Of this oxidized carbon, 45 percent remains in the atmosphere for an average period of about 100 years, and 25 percent remains there permanently. More than that will not be absorbed by the oceans or by biomass in thousands of years. Thus, even a temporary increase in carbon extraction will permanently increase the average temperature on our planet.

But what does "reducing carbon extraction" mean? Does it mean that we should extract the carbon available in the Earth's crust more slowly, or that we should leave some of the stock underground forever, sealing it so that not even future generations can touch it? As was pointed out in chapter 4, a convincing case for sealing off the deposits can't be made, because that would require knowing what people will be willing to pay for the last remnants of carbon and what the costs of extracting those remnants and the associated additional climate damage will be. Such knowledge is simply not possible to acquire at this time. It is entirely possible that the economic utility of the last remnants will be so high that humanity will prefer to extract them despite the climate damage it would cause. Sealing off some of the deposits therefore can't be recommended, aside from the practical impossibility of committing future generations to binding constraints on the use of the world's fossil-fuel resources.

The case for slowing the pace of extraction can indeed be made, however, because of the negative external effect exerted by the greenhouse effect (and of course the present generation can commit to such a policy by just abstaining from extraction). This was shown in chapter 4 in the sections titled "The social norm" and "Why extracting more slowly makes the cake bigger." (The casual reader is strongly advised to read those sections.) It is true that the expected appreciation of the fossil carbon *in situ* constitutes a motive for resource owners not to hasten extraction and instead to safeguard their supplies for future periods. But that only expresses the private advantage of being parsimonious. The social advantage of a lower CO_2 content in the atmosphere in terms of reducing the greenhouse effect is missing from the private calculations. Thus, market forces tend to encourage extracting the fossil fuels too quickly. If today's energy prices were higher than what the market currently settles on, and if they were to increase more slowly with the passage of time, fossil carbon would be extracted more slowly, and that would be better for future and possibly even present generations. If the slower extraction is compensated with a lower rate of accumulation of man-made capital, the standard of living of future generations can be raised without lowering the present standard of living; or, conversely, the standard of living of present generations can be raised without lowering that of future generations.

The resource-conservation strategy doesn't necessarily imply sacrifices by the present generation to benefit future generations. The rationale for resource conservation given in this book is based not on ethics (see the section titled "Nirvana ethics" in chapter 4), but on the hard economics of efficiency and Pareto optimality. For future generations, resource conservation generates the double advantage of reducing future climate damages and increasing carbon consumption. Thus, a compensating reduction in the bequest of man-made capital that just leaves the standard of living of future generations unchanged would yield a higher standard of living for today's generation. The greenhouse effect therefore calls for "green" policies that succeed in slowing resource extraction and in flattening the price path and the extraction path.

Slowing down extraction is *a fortiori* advisable insofar as insecure property rights also imply excessively rapid extraction, as was shown in the section titled "The fear of a coup" in chapter 4. The market for crude oil and (to some extent) the market for natural gas suffer particularly from the frenetic actions of sheiks and potentates whose fear of political upheavals that might provoke changes in resource ownership prompts them to extract their resources much more rapidly than is healthy for humanity at large.

If the greenhouse effect didn't exist and the property rights of the resource owners were secure over the long term, we could simply rely on the market to determine the speed of resource extraction. Today's generations would bequeath later generations a well-composed mix of man-made capital above ground and natural capital in the form of resources below ground. But in view of the two distortions, today's generations are in fact bequeathing a badly composed wealth portfolio, with too little natural capital below ground in relation to man-made capital above ground, that imperils the standard of living of present and/or future generations. One can only strongly advise policy makers to strive for a more ecological portfolio mix.

That is easier said than done. The many well-intentioned attempts by political decision makers appear confused and inconsistent. There are taxes on fossil fuels, but those are encumbered with a plethora of exceptions and with many hundreds of regulations, which muddle the picture. A wild proliferation of legal clauses has taken possession of ecological issues. The law of one price—according to which the incentives to curb CO_2 emissions should be the same everywhere, regardless of their origin, in order to achieve the reduction goals with the least burden to the economy and people—is blatantly disregarded by political decision makers. Western societies are paying far too dearly for the attainment of their reduction goals. The law of one price is the fundamental law of economics. Again, it has nothing to do with justice or ethics; it has to do with the fundamental condition for using resources efficiently to fulfill humanity's goals instead of squandering them. As was argued in chapter 2, the European Union's cap-and-trade system indeed would be an efficient regulatory system if it were equally applied to all sectors of the economy, because it determines the one uniform price of CO_2 emissions. In fact, however, it is applied only to electric power generation and a few other sectors of the economy.

Because they don't harmonize with the cap-and-trade system that already covers all electricity production in Europe, the EU's big support programs for "green" electricity, including subsidies for bioelectricity and feed-in tariffs supporting solar power and wind turbines, have become expensive flops. (See the section titled "Feed-in tariffs, instrumental goals, and the European policy chaos" in chapter 2.) National subsidies for electricity from renewable sources like these are entirely useless, as the overall amount of CO_2 emissions is determined by the cap alone. The subsidies reduce the price of emissions allowances to the point where one country's saving of fossil energy is completely neutralized by the additional use of fossil fuels triggered by this price reduction in other countries. The subsidies merely distort the allocation of "green" and coal-fired power stations across the EU countries, thereby increasing the overall cost of producing electricity.

And, unfortunately for those countries that spend a lot of money subsidizing "green" electricity, the emissions certificates that they can sell to other countries will be counted, for the purposes of the Kyoto Protocol, as if they had been used at home. The wind turbines and photovoltaic panels that came into existence because of national subsidies or feed-in tariffs neither help the climate nor improve a country's Kyoto balance.

Powering cars with biofuels is a nice enough dream of "green" politicians and of farmers in Western countries. However, the fact that producing such fuels requires enormous amounts of land would have catastrophic effects on the world's food prices, as was shown in chapter 3. A policy aimed at putting in the tank what other people would prefer to have on their tables, speeding rather than slowing the inevitable coupling of the oil and food markets that lies ahead, is inhuman and endangers world peace. The Tortilla Crisis was just an alarm bell. Unless it is strictly limited to biogenic waste, the biofuel strategy is not a sustainable option for battling climate change.

The greatest policy challenge of all, however, is finding a way to induce resource owners to modify their supply behavior. If the resource owners stubbornly stick to their extraction paths regardless of how the demand policies of the Kyoto countries affect their prices, those policies will have no effect on the speed of global warming, because at falling resource prices other countries will consume exactly the same quantities that the Kyoto countries manage to relinquish. So far, environmental policy hasn't made the slightest attempt to get the resource owners on board. Policy makers haven't thought about whether their behavior is relevant to the success of environmental policies, let alone how it might be changed. The Western countries spend hundreds of billions of dollars on demand-reducing policies, yet not a word is uttered on the question of whether such policies could affect the supply of fossil carbon, and if so by which mechanism.

Thus, the verdict of this book, up to this point, is that the current environmental policies of many countries are expensive, inefficient, inhumane, and in many cases entirely useless. At best, they haven't been thought through. They simply ignore the supply side of the market.

Even this verdict may in fact be a euphemism for something even more disquieting. There are reasons to believe that the "green" policies may themselves have accelerated global warming. The resource owners didn't stubbornly stick to their extraction paths despite the "green" policies; they moved them in an unexpected direction.

The Green Paradox

Resource owners aren't stupid. Even though the policy measures chosen thus far have had only limited success in fighting global warming, Arab oil sheikhs, Russian gas oligarchs, and coal barons all have realized that a revolution in the world's energy mix is underway. The 1978 Geneva conference of the World Meteorological Organization, the 1988 World Climate Conference in Toronto, the founding of the Inter-governmental Panel on Climate Change in 1988, the 1990 World Climate Conference in Geneva, the Kyoto Protocol (1997), the founding of Green parties in Europe in the 1980s, the rise of Greenpeace in the same decade, Al Gore's environmental activism since 2001, and the Stern Review (2006) didn't go unnoticed. And of course the discussions that led to the establishment of the European Union's emissions trading system, the EU's firm decision to curb Europe's CO_2 emissions by 20 percent until 2020, and the Group of Eight's goal of reducing emissions by as much as 80 percent in a worldwide post-Kyoto agreement to prevent the world's average temperature from increasing by more than 2°C relative to pre-industrial times have also been acknowledged with increasing watchfulness.

The resource owners must also be worried that even in the United States the mood is changing toward a reduction of the demand for fossil fuels. Though the US certainly hasn't yet turned "green" in the European sense, it has attempted to reduce its dependence on fossil fuels by building up an infrastructure for the promotion of bioethanol, which now gobbles up 40 percent of the corn grown in the US. Not even the most optimistic resource owner could overlook the active preparations by the industrial powerhouses of the West—their biggest customers—to develop replacement technologies, or the other measures they are taking to reduce their demand for fossil fuels in the decades ahead.

The policy measures taken and the verbal announcements accompanying them have sent danger signals to the resource owners. If taken literally, they would result in the destruction of future markets for fossil carbon. The announcements must have shocked the resource owners and induced them to revise their extraction strategies. Why leave resources below ground if the probability that they will be sellable at good prices in the future—the basic motivation for private conservation—is diminishing with each year that alternative energy development gains momentum? Under these circumstances, isn't it better to extract much more in the present, and to safeguard the wealth in financial and real assets held in tax havens, than to leave the resources underground and remain at the mercy of alternative-energy freaks and "green" politicians?

What the resource owners heard sounded to them like saber-rattling by people who planned to destroy their markets, or like an announced expropriation of their fossil-fuel deposits. They reacted the same way they deal with the risk of expropriation by rivals who threaten to seize power and reallocate the property rights: by speeding up extraction. They expanded their production capacities and increased their extraction rates in order to sell their resources before it was too late. As a result, more fossil carbon was brought to the markets at falling prices, was burned, and entered the atmosphere. The mere announcement of intentions to fight global warming made the world warm even faster. That is the Green Paradox.[1]

The Green Paradox is theoretical, but behind it lurks an empirical puzzle: the apparent contradiction between huge efforts to curb CO_2 emissions and the inability to even reduce the increase in CO_2 concentration in the atmosphere. As figures 4.1 and 4.2 show, the EU's efforts to curb its carbon consumption didn't even produce a kink in the curve depicting the relentlessly growing trend of worldwide CO_2 emissions. Carbon leakage means that the carbon not consumed in one country is instead consumed in another country if the carbon supply doesn't fall in the same measure as demand was reduced in the countries implementing "green" policies. The Green Paradox could, however, be the deeper explanation. Not only did the carbon supply fail to decline relative to trend as a reaction to reduction in demand; it *increased* relative to trend, because the current reduction, including the public debate accompanying it, was seen as a sign of much more radical demand reductions to come.

In the last section of chapter 4, it was argued that the differences in the price trends for crude oil, natural gas, and coal can be attributed to different expropriation probabilities affecting the owners of the respective resources. An interesting aspect of figure 4.13 is the uniformly flat trend in the 1980s and the 1990s. However, the trend was flat only in nominal terms. In real terms—that is, with inflation subtracted—the trend actually declined over those two decades. Figure 5.1 depicts the same curves adjusted for the US consumer price index. In real terms, all three energy prices declined from the time of the second oil crisis (1980) to about 2000.

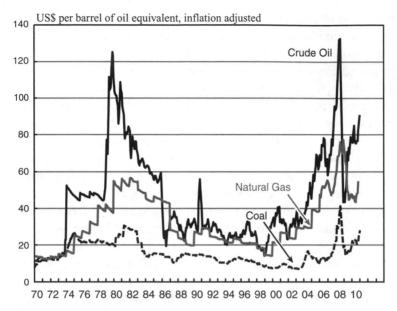

Figure 5.1
Real (inflation-adjusted) prices of fossil-fuel resources. The curves represent the real, inflation-adjusted prices of crude oil, natural gas, and hard coal in energy equivalents with 2009 as a baseline, i.e., the hypothetical prices if the US consumer prices had always remained on their 2009 level. The average annual inflation rate over the period from 1970 to October 2010 was 4.4 percent, which implies that the real growth rates of the resource prices quoted in the last section of chapter 4 are 5.0 percent for crude oil, 3.5 percent for natural gas, and 1.6 percent for hard coal, while the real interest rate on ten-year US government bonds was 2.7 percent. Over the period 1980–2000, during which the real resource prices even declined, the average inflation rate was 3.8 percent. In January 1970, the real price of a barrel of crude oil was $11.16, the price of one energy-equivalent amount of natural gas was $13.65, and the price of one energy-equivalent amount of coal was $8.08. Sources: Hamburgisches WeltWirtschaftsInstitut, *Rohstoffpreisindex* (http://hwwi-rohindex.org); Statistik der Kohlenwirtschaft e.V., *Entwicklung ausgewählter Energiepreise* (www .kohlenstatistik.de/home.htm); Deutsche Bundesbank, *Statistik*, *Devisenkurse* (http://www.bundesbank.de); US Bureau of Labor Statistics, *Consumer Price Index* (http://data.bls.gov).

The period of real price declines coincided with the emergence of the "green" movement and the re-orientation of the world energy policies by way of inducing direct demand restraints and introducing incentive systems to foster the development of "green" replacement technologies. Thus, the threat of market destruction may indeed have increased the supply of fossil fuels enough to more than offset the growing world demand, thereby inducing real energy prices to fall, contrary to what a forward-looking explanation along the lines of Hotelling's theory would have suggested *prima facie*.

The falling real energy prices could, in principle, also be explained by repeated increases in the expected probability of direct expropriation, which would also have led to successive supply increases. After all, a number of oil-exporting countries experienced war and turmoil during the period in question. This could well explain part of the decline in the prices of crude oil and natural gas. However, this explanation makes less sense in the case of coal, which is typically located in safe countries. The real price of hard coal, however, also declined sharply from 1980 to 2000 (by 32.1 percent, a decline of 1.9 percent annually on average). The Green Paradox is an explanation that applies to all three fossil energy sources. After all, all three of them are threatened when politicians try to bring in alternative energies earlier.

The Green Paradox may also explain, by the way, why Germany is the world's largest supplier of lignite (brown coal), with a market share of 18.4 percent, even though it possesses only 1.7 percent of the world's supply. Germany is "greener than green." Arguably there is no country in the world where the "green" movement is as strong as in Germany, and no country that is more determined to take drastic policy measures to curb CO_2 emissions, even at the price of hurting industry and lowering the standard of living. Owning a fossil-fuel resource in such a country is an extremely risky investment, since the government could stop extraction any time. It is little wonder that, under such circumstances, the resource companies operating in Germany, especially the Swedish energy company Vattenfall, are scurrying to secure their wealth before the Greens seize it.[2] More supply of brown coal directly reduces the price of hard coal, as the two fuels are close substitutes in the production of electricity.

Admittedly, all this is not compelling proof that the Green Paradox is the primary force behind the falling energy prices. Future econometric studies may be able to find out how much weight can be given to the alternative threats faced by resource owners. However, it seems clear that the threat of indirect expropriation by "green" policy makers has been added to the threat of direct expropriation by rivals, and that the two threats are exerting a particularly strong downward pressure on fossil-fuel prices and so triggering even more worldwide consumption. Policy makers should at least acknowledge and discuss these effects before naively devising national programs for the development of alternative energies. They should also stop dreaming in public about where the world's energy could come from 30 years from now, and focus instead on coordinating joint effective actions. The falling energy prices are an undisputable and puzzling fact, and the two hypotheses are on the table.

A bit of theory

Though this book cannot go into the mathematics of intertemporal optimization, a bit more precision may be useful to help the reader to fully understand the logic of the argument behind the Green Paradox.[3] Let us consider the carbon market and suppose that in an initial scenario the carbon price and the extraction flow evolve in such a way that the suppliers are indifferent regarding immediate or deferred sales, as described in chapter 4. The expected price increase resulting from an increasing scarcity of carbon is high enough for the appreciation rate of the resources *in situ* to match the interest rate. (Should a risk of open political expropriation also be present, the numerical value of the annual expropriation probability must be added to the interest rate. The nature of the argument, however, isn't altered by this.) This initial scenario for the time path of the resource price is represented by the dark curves in figure 5.2.

Let us now conduct a thought experiment consisting of two steps. We first assume that "green" policies curb demand in certain periods and see how the price path would change if the extraction path or the supply path remained unaltered. Then, in the second step, we analyze how these extraction paths will change as a result of how the resource owners react to the price signals.

The demand policies considered can consist of direct demand reductions caused by simply inducing people to being parsimonious, as well as of indirect reductions by way of subsidizing or enforcing the use of alternative energy sources, as was discussed at the beginning of chapter 4. There are demand curves like figure 4.4 for all points in time, albeit

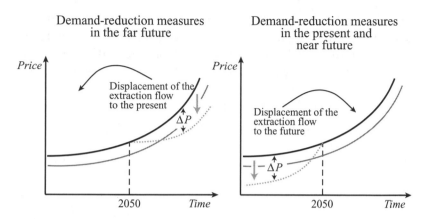

Figure 5.2
When will demand be curbed?

possibly with different positions. The demand curves reflect a continuum of technological replacement possibilities, energy-saving strategies, and "backstop" technologies that become privately profitable at alternative world market prices for fossil fuels. The "green" policies shift these demand curves downward by reducing the critical level of the fossil-fuel price at which the respective replacement technologies that largely explain the slope of these curves become privately profitable. Given the supply, the shift in demand would bring down the price in each period, as the right panel of figure 4.5 shows. But of course supply isn't given.

Two extreme cases are depicted by the dotted curves in figure 5.2. In the first case, demand would be curbed only in the far future, say after the year 2050. Climate forecasts often refer to the times before and after that year. It is also the year on which various G8 and EU declarations about reduction goals focus.

As a result of the curb on demand, given the supply path, the price path falls after 2050 from the solid black to the gray position. The price decrease that results from it at a given time (ΔP) can be called the *price wedge*. Evidently, the price wedge will, in general, be time-dependent.

The new price path can't represent a market equilibrium (in the sense of the Hotelling Rule) in which the resource owners are indifferent between selling now and selling later, as the price will no longer rise continuously in a way that makes the rate of appreciation of the resource stock equal to the market interest rate. Thus, the entire extraction path will change. This is the supply reaction on which this book focuses. If the resource owners know what the future holds in store for them, they will lose interest in selling after 2050, preferring to exploit their deposits before that date. They will keep exploring for new deposits now and in the near future, and will start extraction as soon as they can. In reality the technical preparations will take a few years, but this time span is negligible relative to the time periods considered. In exchange, they will of course have to reduce supply after 2050 in comparison to the original extraction path; after all, they can sell their stocks only once.

Extracting earlier has the consequence of reducing prices much earlier, as a result of increased supply, and not just after 2050, when demand will be curtailed. Conversely, it also means that, owing to curtailed supply, the prices after 2050 relative to the gray portion of the price path will rise somewhat, or will not fall so markedly relative to the black path as would have been the case if extraction hadn't come earlier. This stretches the expected price path again, shifting it to the gray position, which lies everywhere below the original one, which would apply in absence of a "green" demand policy. Although the price in each period is lower than before, the resource owners are again indifferent between extracting now and extracting in the future. The expected appreciation of the resource *in situ* again equals the market interest rate.

Unfortunately, bringing extraction forward to the period before 2050 means that some of the carbon deposits will be burned earlier, reinforcing the greenhouse effect. This is the Green Paradox: Announcing a future reduction in the demand for fossil fuels increases the supply in the present and speeds up global warming.

Let us now consider the other case, illustrated by the right panel of figure 5.2. Demand will be reduced relative to what it otherwise would have been only in the near future until 2050, but evolves afterward as in the thought experiment's initial scenario. With the old supply path, prices would decrease as shown on the gray path. The consequence is a departure from the old supply path because it isn't profitable for the resource owners to offer their products before 2050. In reality, deposits already being exploited might continue to be exploited, but no new fields would be brought into production; as a result, supply would be reduced relative to the old extraction path up to 2050 and increased after that year. This reduces prices after 2050 relative to the old price path and increases them a bit before 2050 relative to the gray path. The new stretched price path, which satisfies the Hotelling Rule, is shown by the gray curve. It runs everywhere below the original one, which would have prevailed without the government's demand-reducing measures. The deferral of extraction embodied by the new price path is exactly what we need in order to slow down global warming. Earth stays cooler for the time being.

The conclusion to be drawn from these considerations is that a curb that affects demand uniformly along the old extraction path in all periods before and after 2050, so that the "pricing pressure" is everywhere the same, provokes no supply reactions at any time. If future demand is curbed, fossil fuels will be extracted more intensively in the present; if present demand is curbed, they will be extracted more intensively in the future. If demand is curbed uniformly in the sense of exerting a constant pricing pressure in all periods, the extraction path will remain unaltered. It is almost like a hydraulic system in which various pistons are interconnected through a liquid-filled pipe system. If one piston is pushed, the others move up; if all the pistons are pushed simultaneously with equal force, none moves.

The precise condition for the neutrality of demand policies with regard to supply reactions can be defined in terms of the time dependence of the price wedge that the demand policy exerts, given the old extraction path, that is, before resource owners react. The pricing pressure is uniform in the sense described above precisely when the price wedge caused by "green" policies, with the extraction path that would have prevailed in the absence of such policies, grows at a rate equal to the market rate of interest.[4] Finance experts would say that the value of ΔP discounted to the present—the so-called present value of ΔP—remains constant over time. If the pricing pressure exerted by the policy measures of the Kyoto countries is uniform in this sense and is never so large that it pushes the price below the unit extraction cost (an aspect that will be investigated below), the "green" policies will not bring about supply reactions at any time. The price-inelastic supply introduced for didactic purposes in chapter 4 (figure 4.5, right panel) thus proves to be a possible, albeit special, case in the middle rather than at the edge of the spectrum of possibilities.

This insight allows us to cast a new glance at the "green" demand-reducing policies of the Western world. Evidently, if these policies want to defer extraction to the future, they must curb demand more strongly in the present than in the future, in such a manner that the present value of the price wedge declines with the passage of time—in other words, in such a manner that the price wedge, given the old extraction path, increases at a rate below the market rate of interest. Only then will climate warming be slowed down as a result of a postponing of extraction. Policy, therefore, must be very "green" at the beginning and then pale gradually over time.

Exactly the opposite is happening now. Policy has become increasingly "greener," and so the pricing pressure has been increasing. In Europe and in other parts of the world, laws and regulations have come in quick succession, each improving or expanding previous ones, raising the expectation of further tightening in the future. All of this must have strengthened the resource owners' concerns that in the future their possibilities for exploiting their fossil carbon still *in situ* would be limited and must therefore have induced even more rapid extraction, as the Green Paradox predicts.

The Green Paradox and carbon leakage

The problem of carbon leakage was examined in chapter 4. The "Kyoto countries" implement "green" demand-reducing policies, while the "non-Kyoto countries" abstain from such policies. It was shown that, given the supply of fossil carbon, the demand-reducing measures of the Kyoto countries depress the world market price until the quantities set free by such countries are fully absorbed by the non-Kyoto countries. The falling price induces the Kyoto countries to themselves compensate one part of the demand reduction induced by their policy measures. The other part is compensated by the countries that don't follow any "green" policies. The upshot is that carbon leakage is 100 percent—that is, the entire net reduction in demand in the Kyoto countries, induced by the "green" policies, is compensated for by an increase in demand in the non-Kyoto countries via an appropriate decrease in the world market price of carbon.

The analysis, which centered on figure 4.6, applied to a certain period. As supply was taken to be exogenous, the carbon price was determined solely from the interactions of both buyer groups as they compete for the available flow of carbon in the market. The flow itself, i.e., the supply, was represented by the width of the chart (the length of the dashed line at the bottom), and it was shown that "green" policies cause a decrease in the world market price of carbon of a ΔP magnitude. This ΔP is in principle the same price wedge as that shown in figure 5.2 and to which the neutrality rule derived above applies. But whereas figure 5.2 shows the intertemporal evolution of the price wedge, figure 4.6 showed how a given period's price wedge develops in the world markets if only one group of countries carries out "green" policies. It is now time to combine the international analysis of carbon leakage and the intertemporal analysis of resource supply. To do this, we will again resort to a thought experiment.

There are many periods and therefore many charts of the type shown in figure 4.6. They differ in the position of the period-specific demand curves of the Kyoto and non-Kyoto countries, in the latter case with regard to the demand curve both in the absence of "green" policies and in the presence of such policies. Each graph has a different width, corresponding to the supply offered by the resource owners during the period in question. In the first step of our thought experiment, this width doesn't react to a change in the world market price of carbon. Thus, in each period, an idiosyncratic ΔP resulting from the "green" policies then in force is determined.

If the borderline condition for a neutral "green" policy explained above holds (that is, if the pricing pressure stays constant over time), the assumption of a non-reaction of supply would hold even if we were to move to the second step of the thought experiment and allow for an endogenous change in supply. This would be the case if the present value of ΔP is the same in all periods—that is, if ΔP rises at a rate that just happens to equal the capital-market interest rate. Though this is a possible case in the middle of the spectrum of possibilities, useful for didactical purposes, it is of course a special and unlikely case.

If the "green" policies become tighter over time, and particularly if they are announced now, well ahead of being carried out, such that the present value of ΔP increases with the passage of time, the condition for the Green Paradox is satisfied.[5] The producers react by selling their resources earlier, so the charts grow wider in some initial periods and become narrower in later periods. The announced increase of the pricing pressure by way of strengthening of demand-reducing policy measures in the Kyoto countries causes the resource owners to bring their sales forward to such an extent that in the initial periods the world market price of carbon falls so much that the non-Kyoto countries' demand increases by more than the Kyoto countries' demand shrinks. Perhaps the Kyoto countries' own demand increases too, because the effect of a falling world market price of fossil fuels more than offsets the effects of the demand-reducing policies. In any case, aggregate worldwide demand increases because in a market equilibrium someone will buy the additional quantities supplied—a truly frustrating result.

The reader may wonder whether this scenario is not too gloomy. At least the countervailing demand increases of the Kyoto countries resulting from the fall in the world market price of fossil fuels could be blocked by resorting to a policy of strict quantity constraints, such as the European cap-and-trade system. Wouldn't such a system, which determines the aggregate quantities without having to resort to price signals, instead setting a rigid cap on the number of emissions certificates issued, be more successful in curbing worldwide demand, and avoid the Green Paradox?

The answer is No. The reason is illustrated in figure 5.3, which follows the spirit of figure 4.6 but which depicts the "green" policy of the Kyoto countries in the form of a quantity constraint, the cap being set by the emissions trading system. The cap equals the distance between point B'

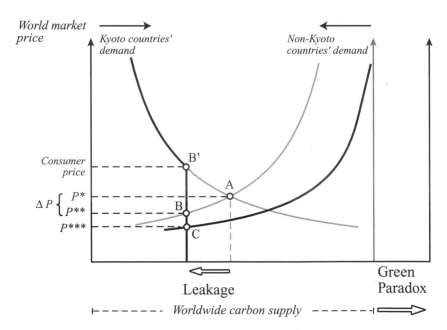

Figure 5.3
Carbon leakage and the reaction of resource owners' supply.

and the left vertical line. The Kyoto countries' carbon demand curve therefore sprouts a vertical branch leading from B' downward. No matter how much the world market price of fossil carbon falls because of the demand restraint, the Kyoto countries can't demand more carbon than the amount implied by the cap.

Without emissions trading, the world market equilibrium would be represented by point A. The distance between A and the left vertical line would represent the carbon consumption of the Kyoto countries, and the distance between A and the gray, inner vertical line on the right would represent the demand of the non-Kyoto countries. The corresponding world market price would be P^*.

With the demand restraint enforced by the cap, the world market price falls to P^{**} as the world market equilibrium moves from A to B. Obviously, the new equilibrium implies a shift of carbon consumption from Kyoto countries to non-Kyoto countries, as depicted by the left-pointing arrow underneath the horizontal axis. There is 100 percent leakage of the carbon saved in the Kyoto countries if the aggregate supply (as measured by the distance between the left vertical line and the gray, inner vertical line on the right) is constant.

Consumers in the Kyoto countries now pay two prices: the world market price P^{**} and the price of the emissions certificates. The latter equals the distance from B to B'. It can't be lower than BB', because then there would be an excess demand for certificates, pushing their price up. And it can't be higher, because then the demand for certificates would be lower than the outstanding stock, forcing the price to decrease. Note that although the world market price of carbon is lower than what it would have been without the cap-and-trade system (point A), the sum of the two prices is higher. Domestic energy prices rise because of the cap-and-trade system, while the world market price decreases because of that system. If domestic energy prices didn't rise, the Kyoto countries' consumers wouldn't demand less carbon, which is the precondition for the decrease in world market prices.

Emissions trading exerts pricing pressure, and once again the actual supply reactions depend on how this pressure changes over time, the pressure being measured by the present value of ΔP that results from the cap-and-trade system, with the old extraction path. The pricing pressure is stronger the fewer allowances are allotted, because this constraint reduces the demand for carbon on the part of the Kyoto countries. If the pricing pressure remains constant, the extraction-and-supply path shows no reaction. If it decreases, supply will shift to the future. If it increases, extraction speeds up, giving rise to the Green Paradox.

This last case is particularly plausible. The European Union has already reduced the number of allowances it allots by 5 percent from the first trading phase (2005–2007) to the second (2008–2012), and announced that the number will be reduced by 21 percent by 2020 relative to the first trading phase.[6] Furthermore, third-phase allowance certificates will have an automatic expiration date, which will reduce their number unless new allowances continue to be issued. The UN trading system, discussed in chapter 2, also contemplates reducing the number of certificates allotted in relation to the size of the participating countries. Moreover, as was explained in chapter 2, the G8 countries have already committed to the goal of reducing their carbon output by at least 50 percent, if not 80 percent, by 2050.

The increasing imposition of constraints on a subset of the carbon-consuming countries means that the vertical branch of the demand curve resulting from emissions trading will move steadily to the left, even though continuing economic growth actually pushes the demand curves of both groups of countries steadily upward or to the center of figure 5.3. The constraints represented by emissions trading will become increasingly tighter, perhaps enough to result in a growing present value of ΔP. This would then result in the Green Paradox, that is, a speeding up of extraction and an increase in the current and near-future carbon supply. This again means that the width of the graph shown in figure 5.3 widens in some initial periods and shrinks thereafter (always, of course, relative to what it otherwise would have been in the same period). Figure 5.3 shows this reaction in one of the initial periods. The right-hand vertical boundary line and the non-Kyoto countries' carbon demand curve together move to the right, taking up the respective solidly drawn positions depicted, as more carbon is supplied. It follows that the world market equilibrium moves from B to C, resulting in a further price decline that again stimulates the non-Kyoto countries' demand just enough to absorb all the extra supply coming from the resource owners because of the Green Paradox.

This is a veritable environmental policy catastrophe. For one thing, the carbon not consumed by the Kyoto countries migrates to the United States, China, and all the other countries that aren't constrained by the Kyoto Protocol. For another, these countries will, in addition, consume all the extra supply that the resource owners bring to the market as they grow afraid of the gradual destruction of their markets through tightening of the cap. In a sense, this is more than a 100-percent carbon leakage from the Kyoto countries to other countries. It could hardly be worse.

In the context of the metaphor used in chapter 4, it would be like this: The EU and the other "green" countries leave the church after depositing their alms in the box. Behind them come the Chinese, the Indians, and the Americans. Not only do they dip into the collection box and take the money deposited by the EU and the other "green" countries; they also plunder the treasury of the church.

Will production costs and replacement technologies stop extraction?

One reason why the Green Paradox is disturbing is that the abnormal supply reaction of the resource owners contradicts all experiences with normal markets. In a normal market, the price selects, from the distribution of unit costs across actual and potential firms, those companies that are profitable and merit existence. The higher the price, the larger the number of potential firms joining the market because their unit costs are below the price and lead them to produce. Conversely, a price decrease wipes out those firms whose unit costs now lie above the price, reducing the market supply. The marginal firm that just survives has unit costs approximately equal to the market price.

Applying this basic economic logic to the market for fossil fuels would suggest that those fossil-fuel deposits whose extraction costs exceed the product market price wouldn't be exploited today, but those with unit costs below the market price certainly would. Since the least profitable deposits being extracted today have extraction costs barely below the market price, a price decrease would make them unprofitable and drive them out of the market, reducing the supply of fossil fuels.

But the analogy is entirely wrong, because extraction costs of exhaustible natural resources aren't the same as production costs of reproducible goods. Unlike the latter, the former are only loosely related to price. The price of an extracted resource is far above the extraction cost, and hence limited price decreases such as those carried out by environmental policies wouldn't drive such deposits out of the market for cost reasons. While nothing is being extracted from deposits whose unit costs exceed the product market price, the opposite isn't true. Reserves are resources that promise profits if extracted today; nevertheless, they are left in the ground for later use, with only tiny shares being taken year by year. As was shown in figure 2.6, the reserves suffice for quite a bit of time yet. The running time or reserves-to-production ratio is 41 years in the case of oil, 60 years for natural gas, 137 years for anthracite coal, and 293 years for brown coal (lignite).

The fact that many resource deposits aren't being exploited even though they would be profitable is attributable to their owners' expecting even greater profits in the future as a result of rising world market prices. In a market equilibrium, the expected future profits generated by deferring extraction are high enough to compensate for the loss of interest income resulting from a delayed cash flow, and normally the supply in all periods is kept sufficiently scarce to generate huge profit margins over the extraction costs.

Resource companies, in a sense, even try to maximize the difference between price and extraction costs by exploiting the low-cost deposits first. Indeed, if they want to maximize their profits and wealth, they should refrain from extracting from high-cost deposits until the low-cost ones are exhausted, because the money thus saved can be invested in the capital market and generate an interest income.[7]

This logic applies fully to deposits owned by a single company that is free to choose the order of extraction from deposits of different accessibility. When these deposits are owned by two or more companies, perhaps located in different countries, things can be a bit different. A company that owns a high-cost field and needs money but is borrowing-constrained may find itself tempted to extract its resources prematurely. However, even such a company could offer the owner of low-cost deposits a mutually beneficial bargain that results in exhausting the low-cost deposit before extraction from the high-cost deposit begins. For example, it could offer to exchange some of its high-cost resources against unused low-cost reserves from another company, offering a side payment in terms of deposits given in exchange to compensate for the higher extraction cost. This would enable the original high-cost company to extract from the low-cost field and give the original owner of that deposit the opportunity to invest the money received in the capital market and earn interest on it. As this interest comes in addition to the compensation for the cost disadvantage, it is a true gain resulting from the improvement in the order of extraction. If both parties find a way to share this interest income, they would both be better off. Admittedly, political constraints may prevent the companies from carrying out such deals, and, as discussed above, companies in countries with uncertain property rights may feel strong incentives to extract prematurely. However, as a rule, markets will tend to extract the low-cost deposits first; therefore, there will typically be a substantial profit margin.

The economics of resource extraction also stand the usual logic of marginal costs on its head. With normal goods, marginal costs are the highest production unit costs across all sites, as sites with lower unit costs operate at capacity and more output will only be available when new firms with higher costs enter the market. In the case of exhaustible resources, things are the other way around. Here the marginal costs are given not by the highest unit costs but by the lowest, because these are the additional costs that result from additional extraction.

All of this shows that the price of renewable resources has little connection with extraction costs.[8] It is rather a pure scarcity price, similar to the price of a plot of land. Extending the example of price formation for old Rembrandts given in chapter 4, we might say that the price of fossil fuels is as much determined by the extraction cost as the price of a Rembrandt is determined by the cost of the master's paint.

Table 5.1 shows how unimportant extraction costs are.[9] The second column gives the range of unit extraction and exploration costs relative to the world market price of the respective fossil energy source; the third column gives the corresponding weighted average of the actual cost distribution across the different deposits, the weights being given by the current production shares of these deposits. In the case of coal, the extraction costs are significant; nevertheless, with average unit costs at 45 percent of the price and the upper range at only 61 percent, that is still far from what cost-driven pricing would have implied. To explore for and extract crude oil, only 17 percent of the oil price is needed on average, and for natural gas 16 percent would suffice. Though in absolute terms the mean price of crude oil in the period 2005–2009 was $69 per barrel, only $11.4 per barrel was needed in an average field to cover exploration, extraction, and readying for shipping. Arab oil producers had extraction costs of just $2 per barrel. If the arbitrage between high-cost and low-cost reserves worked properly, these would be today's marginal cost of resource extraction.

Of course table 5.1 is just a snapshot. Even though markets deviate somewhat from the optimal order of extraction, the easier fields will be exhausted first, and resource owners will have to resort to increasingly difficult fields with ever-higher extraction costs. This tends to increase the cost-price ratio. However, in a market equilibrium that generates

Table 5.1
Unit extraction and exploration costs relative to world market prices (averages 2005–2009). Source: H.-D. Karl, "Förderkosten für Energierohstoffe," *ifo Schnelldienst* 63 (2010), no. 2: 21–29.

	Range	Production-weighted average
Crude oil	3%–67%	17%
Coal	28%–61%	45%
Natural gas	4%–35%	16%

enough incentives for leaving some of the resource below ground for later use, the quantities extracted become smaller and smaller relative to aggregate demand. This increases the price, which in itself tends to reduce the cost-price ratio. It isn't clear what the net effect of these two forces is, because it depends on subtle characteristics of the cost distribution across the deposits and the shape of the world's resource demand curve. However, it is a standard result of most resource-extraction models that the unit extraction costs will never be able to catch up with the rising resource price, so that a cost-driven cut-off point for extraction will never occur.[10]

Many people believe that this is no longer true when there are replacement ("backstop") technologies for fossil fuels, arguing that such technologies will automatically impose a price ceiling on fossil fuels, putting a halt to extraction when the extraction costs tend to surpass this ceiling. Some even believe that this event is very near and that within a few years solar power will be able to meet all our energy needs and will crowd fossil fuels out of the market. These views are not well founded.

For one thing, table 5.1 provides little support for a quick cost-driven crowding-out process. Even if solar power were able in the near future to provide electricity at or below today's cost of electricity from fossil sources, which would require cutting costs to one-fifth or even one-eighth of the current level,[11] that would not bring a cost-driven end to fossil energy, as the prices of fossil fuels would have to fall quite radically before even coming anywhere near the extraction costs. After all, in the period 2005–2009 they exceeded the average extraction costs by a factor of 6 in the case of crude oil and natural gas, and by a factor of 2 in the case of coal. Before the Arab states would put an end to oil extraction for cost reasons, the price of oil would have to go from the $120 per barrel prevailing at this writing, in April 2011, to the neighborhood of

$3—that is, to one-fortieth of its current level. It is difficult to imagine something like this coming to pass, no matter what technological breakthroughs are expected.

For another, the existence of replacement technologies doesn't in itself imply a price ceiling that could preclude deposits from being exploited for cost reasons. Of course, with increasing prices more and more replacement technologies will become profitable and will successively crowd out fossil fuels from more and more uses, but this isn't the same thing as a price ceiling. In fact, all the considerations in this chapter and the preceding one were explicitly built on the idea of more and more replacement technologies' coming in at higher prices. (Recall figure 4.4.) Despite the existence of such replacement technologies, it is possible, if not likely, that with increasing scarcity the price will always increase enough to stay above the rising unit extraction costs.

And, of course, replacement technologies don't contradict the Green Paradox in any sense. To the contrary, this paradox comes about because public policy measures usher in replacement technologies earlier than the market would have done, causing the increasing pricing pressure that triggers the paradox.

Nevertheless, it is possible that, in contrast with what has been assumed thus far, the price pressure exerted by "green" policies will push the price below the unit extraction cost for some of the resource stocks. In this case these policies will reduce the (economically) exhaustible stock. This will in itself tend to reduce the supply in all periods, including the present, and work against the Green Paradox. It will not, however, necessarily exclude the paradox.

The issue can best be understood by recalling the neutrality condition for a non-reacting supply, which requires a uniform pricing pressure in the sense that the price wedge created by "green" policies is the same in all periods in present-value terms. If a uniform pricing pressure drives the price below the unit extraction cost for some high-cost deposits at some future time, resource owners will have no direct incentive to change the intertemporal pattern of extraction. However, as the exhaustible resource stock falls, they will sell less in all periods. Thus, the uniform pricing pressure no longer is a neutrality condition.

But what if the pricing pressure isn't uniform but increases over time? In this case, the supply reaction is no longer clear, because there are two countervailing forces. On the one hand, resource owners extract the exhaustible stock faster; on the other, this stock itself is reduced. The net result of these two countervailing forces depends on how rapidly the "green" policies bite, and on the size of the high-end resource stocks that are being cut off by the pricing pressure. If relatively few stocks are cut off, and/or if the "green" policies become draconian relatively quickly, the Green Paradox will occur nevertheless. This could be the case if there were a large gap between the extraction costs of conventional and non-conventional resources, and if the replacement technologies ruled out exploitation of the non-conventional resources even before "green" policies were pursued. In that case it would easily be possible for the pricing pressure brought about by "green" policies to exert only a small effect on the exhaustible resource stock, so that more or less the usual conditions for the Green Paradox would still apply.

Yet another theoretically conceivable case is that of a *perfect* substitute technology for fossil fuels that imposes a hard price ceiling. If the "green" policies lower this ceiling, say by subsidizing the substitute technology, there are again countervailing forces on the current supply behavior of resource owners. On the one hand, the lowering of such a ceiling is an extremely strong incentive to shift supply to the present; on the other, it may exclude many deposits from being profitably exploitable even in the long run. A number of authors have reacted to my work by tackling formally the case of perfect substitutes in theoretical models. The models unanimously show that the Green Paradox remains possible despite the assumption of a hard price ceiling, but the conditions under which it will or will not prevail are rather complicated and special. Interested readers are referred to the literature.[12]

How likely is the existence of a *perfect* substitute for fossil fuels that limits the exhaustible resource stock? In the section titled "More than electricity" in chapter 3, where this topic was discussed in detail, it was argued that only biofuels could be a nearly perfect substitute for fossil fuels. However, this substitute isn't available at a stable cost and therefore can't impose a fixed price ceiling. Owing to the enormous amount of land required and the increasing rivalry with food production, increasing quantities of biofuels can become available only at increasing prices. Biofuel therefore doesn't result in a hard price ceiling, which is the unfortunate reason for the Tortilla Crisis examined in chapter 3. It is to be hoped that policy makers will hurry to erect stronger barriers between the markets for biofuels and for fossil fuels, instead of subsidizing their linkage.

A renewable energy substitute that might be seen as a candidate for putting a ceiling on prices is electricity from solar, wind, and hydro power. However, as was explained previously, electricity is a very limited substitute for fossil fuels, since it can't easily be stored. Even the best batteries available today, such as those used in laptop computers, have less than 2 percent as much energy content per unit weight as diesel fuel or kerosene. For transportation, which in the OECD countries consumes about one-third of the final energy, fossil fuel is an unbeatable energy-storage medium. A hydrogen economy could go some way toward replacing the fossil-fuel economy even for transportation, but it would have a hard time providing a perfect substitute for fossil fuels in all their uses, particularly in aviation. For that matter, as long as the Chinese or the American air force is unwilling and unable to run its jet fighters and rockets on hydrogen, the *perfect* substitute for fossil fuels that is needed for the assumption of a hard price ceiling is not in sight. In view of the military uses of fossil fuels, there will probably never be a point in human history at which extraction stops because the resource consumers' willingness to pay doesn't suffice to cover the extraction costs.

Temporary and permanent price changes

The Green Paradox describes a reaction of the fossil carbon supply to price changes that at first glance looks perverse. "Green" policies that reduce the demand for carbon and depress the world market price of carbon end up increasing the carbon supply. The rationale of this statement has been challenged by pointing out that, empirically, fossil-fuel prices and extraction quantities are positively rather than negatively correlated.[13] Wouldn't this positive correlation contradict and exclude the Green Paradox?

It wouldn't, because the Green Paradox doesn't state that current supply reacts inversely to *current* price changes, but rather that it reacts inversely to *future* price changes. A future price decline induced by "green" policies induces a future reduction in supply, and this translates into a current increase in supply, insofar as the resource has to be sold at some point. Conversely, a present price decrease induced by such policies induces a present fall in supply, which results in a future increase. The oil gushes out of the ground in those periods in which the price isn't depressed. It is as simple as that. The seeming paradox arises only when current price changes are accompanied by even bigger future price changes, such that the future price changes dominate the supply reactions.

Of course, such a paradox can't result from cyclical price movements like those occurring over the business cycle, because a cyclical movement by definition is one that isn't accompanied by a change in the trend. If the economy booms and demand increases, this doesn't imply that it will also boom in the future; the same thing applies to a slump and its attendant deterioration in resource prices. Such price changes will always lead to normal (i.e., parallel) supply changes that go in the same direction. However, abnormal supply reactions can result from "green" policy measures, since such measures typically are of a structural nature, extending over longer periods of time.[14]

Paling green

If a policy that becomes "greener" over time prompts resource owners to accelerate extraction of their stocks and thereby increase global warming, it seems appropriate to urge policy makers to adopt the opposite policy: initially pursuing an intensively "green" policy that reduces consumption drastically in the present and near future, then allowing this policy to pale by gradually relaxing the emissions caps. Such a policy would initially exert strong downward pressure on prices, but would later lessen the pressure, with the effect that resource owners would extract less at the beginning and more later. Global warming would slow down, as desired.

Unfortunately, this is only a theoretical solution, because an increasingly less "green" policy would lack credibility among resource owners. All proposals for long-term climate goals put forth by policy makers point in exactly the opposite direction. Constraints are to be small at the beginning but are to increase in the future. The G8 Summit at Toyako in July 2008 (at which the participating countries committed to a 50–80 percent reduction goal by 2050), various proclamations of the EU Commission, and other statements by political bodies follow the same pattern. The largest reduction efforts are to be made in the far future; current generations are largely spared. Politicians cannot do otherwise, alas, as they don't want to inflict the pain of immediate reductions on voters. The year 2050 is so far in the future that the boldest policy proposals for that year can be made today without scaring voters. After all, the onus will fall on future citizens and future politicians. The consequence of this delaying policy is that the resource owners will extract their resources earlier. The quantities that the politicians announce will be restricted in the future are being extracted from the ground all the more copiously today.

An environmental policy that is subject to the constraints of democratic discussion and that limits itself to influencing the demand for fossil fuels can't persuade resource owners that the price of their products will be less affected in the far future than at present or in the near future. On the contrary, the resource owners will be plagued by the fear that, as the planet becomes warmer and the resulting climate damage becomes more apparent, this policy will be tightened even further. As a result, it can hardly be expected that a demand policy that attempts to influence supply through price signals will ever make a contribution toward curbing climate change.

Super-Kyoto

One way to overcome this problem might be through the formation of a seamless consumer cartel in which all consumer countries take part. Demand policies are ineffective if they encompass only some countries, as they will then operate only through price signals. The non-participating countries will gobble up, at lower prices, not only the quantities of fossil fuels that are set free thanks to the efforts of the Kyoto countries, but also the additional quantities that resource suppliers may bring to the market out of fear of future deterioration in the business environment for their products. However, if *all* consumer countries accept a cap on consumption, demand policies may work, as the suppliers will find no takers for their products and will have to reduce extraction. Expectations about the future will no longer play any role. With consumption caps valid for all consumer nations, the playing field will be tilted in a direction that does something for the climate.

The consumption caps could come about through a global certificate trading system that would extend the inter-state trading system introduced by the UN for a number of countries in 2008. Perhaps this system could even be extended along the lines of the European Union's cap-and-trade scheme (in operation since 2005), which allows individual companies to participate in emissions trading. Granted, it would still be a market system that allocates carbon volumes to the individual countries, but now the extraction path would be set not by the resource owners but by the UN. The resource owners wouldn't be able to extricate themselves easily from the grip of the UN.

During the climate summits in Copenhagen in 2009 and Mexico in 2010, some countries, notably the EU countries, tried to convince the world community that such a Super-Kyoto system was needed. But it was to be expected that the proponents would fail again, as the road to an all-encompassing consumption cap is still a long one. As was explained in chapter 2, thus far only the 27 EU countries, Canada, Australia, Iceland, Japan, New Zealand, Norway, Russia, and Ukraine have accepted caps on CO_2 emissions. The rest of the world, responsible for more than 70 percent of CO_2 emissions, has kept quite clear of such commitments. Unfortunately, it is hard to imagine China or India being willing to stifle its economic growth by committing to a cap on its own CO_2 emissions. It is equally hard to imagine the West making the major concessions that would have to be made to get the emerging economies on board.

In principle the Super-Kyoto system would be similar to the rationing that was practiced in many European countries just after World War II, when in order to buy food one needed rationing coupons that were issued by governmental agencies according to social criteria. To buy a pound of butter, one had to pay the shopkeeper the regulated price of the product and, at the same time, give him a butter coupon. If one didn't have enough coupons, one had to trade with other coupon recipients. Extending the use of UN certificates to all countries around the world would be a very similar mechanism. The total amount of carbon available to each country could be rationed this way, and the distribution of the certificates via the UN trading system and subordinate regional trading systems, such as that of the EU, would determine where carbon would be burned.

If mankind wishes to pursue the path to a worldwide cap-and-trade system, it is important that it do so quickly. Any delay is poison for the climate, not only because in the meantime emissions will continue unabated but also (and especially) because piecemeal inclusion of additional countries would exacerbate the Green Paradox even further. If the number of countries accepting caps on their emissions increases only bit by bit, this will give rise to an increasingly larger pricing pressure that will induce the resource owners to counteract the worsening of their profit margins by speeding extraction. Paradoxically, the more successful the world climate summits are in gaining members to the worldwide demand cartel over the coming years, the more rapidly the world's climate will warm in the initial stages. Only taking the resource owners by surprise, with an immediate completion of the consumer cartel that proceeds so rapidly that the resource owners no longer have time to react by accelerating the extraction of their resources, can bring about the desired effects. If this surprise effect is lost, everything may get worse instead of improving.

The hope that the emerging economies—which have managed to stay clear of entering into binding emission cap commitments—will suddenly realize their folly during one of the upcoming climate summits, and make a surprise possible, is unfortunately very slender. There is too much at stake for these countries. But they have to come on board. Many think it would be a success if China and India were to present themselves as model countries in capping emissions, accepting a gradual tightening of constraints over the coming decades. But in fact that would be a disaster rather than a success, because it would set the forces of the Green Paradox in full motion. Realistically, mankind has no more than 10 years to come up with a worldwide emissions trading system. Speed is of the essence.

It isn't quite clear exactly how prices for fossil fuels would form in the new system, because there would be complicated interactions between the consumer cartel and the suppliers. However, the demand constraints would definitely imply that prices that the sellers of the resource receive would be lower than in the case of a *laissez-faire* market solution. From an economic viewpoint, the Super-Kyoto system amounts to a partial expropriation of the resource owners and a partial substitution of the market mechanism by a centrally planned control of quantities. Since a consumer is allowed to use the resources only if he or she can produce the UN rationing coupons, the UN will become, in economic terms, albeit not legally, a co-owner of the fossil fuel. If the UN gives the national governments the right to sell these rationing coupons (as might be the case within the EU for the third trading period starting in 2013), then it will transfer its ownership rights to the national governments. The national governments will then collect some of the funds that consumers otherwise would have given to the resource companies to buy the fossil fuel. The market value of the stocks *in situ* or the stock market value of the companies that own the resources will fall accordingly.

Incidentally, as the carbon burned is the same as the carbon extracted, controlling the overall emissions volume effectively is the same as directly controlling the extraction path of the exhaustible fossil carbon resources: in the end, the whole thing is a central-planning solution for the use of fossil energy, made more palatable only thanks to emissions trading.

Whether we should set out along this path, in light of the negative experience with central planning in communist countries, is a highly complex question that is difficult to decide. In the final analysis, we will probably have no choice but to let the UN take over the central planning, as markets have shown that they fail miserably when it comes to determining the optimal speed of carbon extraction in the presence of the global-warming externality and the risk of expropriations.

As history has shown, choosing a central-planning solution will certainly involve various negative incentive effects. A power center will grow up around the UN that will try to extricate itself from democratic controls. Countries will begin to struggle with one another over who is to be favored in the allocation of the certificates, and will seek to obtain exceptions from the necessity to purchase certificates. This, in turn, will further strengthen the power of the UN bureaucracy. A worldwide black market for carbon might arise, with a Mafia-style counterforce not subject to democratic controls.

The resource-owning countries will do all they can to resist such a solution. They will try to prevent the UN from forming a worldwide demand cartel, and by granting special delivery arrangements for fossil fuels they will try to keep as many countries out of the cartel as they can. They will also try to form a counter-cartel. In light of these developments, it isn't surprising that OPEC is flirting with the idea of admitting Russia. Moreover, the resource-owning countries will attempt to develop their own economies in such a way that they will be able to exploit their own fossil fuels without limitations imposed by the UN. In view of these considerations, Dubai's breathtaking economic development can surely be understood as a rational counter-strategy of a significant resource-owning country.

However, these avoidance maneuvers may induce the consumer countries to develop their own counterstrategies. The countries participating in the cartel will not allow individual countries to acquire fossil carbon without the proof of certificates, and they will build up trade barriers to punish those who deviate. All this will create a considerable conflict potential that could lead to military conflict.

Only the horror of further warming of the atmosphere, in combination with the fact that the consumer countries will have to keep spending considerable parts of their real income for the acquisition of constantly dwindling amounts of carbon, makes the worldwide demand cartel that the UN is planning attractive. Policy makers have the choice between Scylla and Charybdis. Scylla, the monster that regularly ate some of the sailors who dared to cross the Strait of Messina, reminds us of Hobbes's Leviathan, representing the greedy state guzzling tax revenues. Charybdis is a whirlpool large enough to suck in ships. Odysseus opted for Scylla because he wanted to avoid the risk of losing his entire ship and his men in a whirlpool. Scylla may also be the less disastrous option for mankind. Giving up some of its liberal goals may be necessary, for it definitely should avoid going under in the whirlpool, into which the world will navigate if global warming continues relentlessly.

Leading by example?

Installing a Super-Kyoto system is easier said than done, because it is not at all clear what strategies Europe's pushy "green" countries could use to bring on board the Chinese, the Indians, the Americans, and all the others who so far have not accepted binding emissions constraints. The strategic interactions between countries in international negotiations are difficult to predict, but they have to be understood before a negotiation strategy can be developed.

Up to now this book has discussed how other countries and the resource owners react to the Kyoto countries' carrying out of "green" policies. The reaction described was always a private reaction of other market agents. For example, it was argued that the Kyoto countries' reduction in carbon demand brought about by their respective governments' actions led to a decline in the world market price of carbon, which then triggered additional demand in the non-Kyoto countries and unexpected supply reactions. No policy and no collective strategy on the part of the other countries was considered, just individually optimal reactions of market agents that adjusted to the policy measures of the "green" countries, which they had to take as a given.

The interactions that take place at the level of international negotiations between states follow a different logic and should not be confused with the market reactions. For example, it could well be the case that the carbon demand restrictions imposed by governments of the Kyoto countries will lead to similar policy reactions by governments of the non-Kyoto countries. Or perhaps the reverse is true, and others will just lean back and enjoy the fruits of the Kyoto countries' altruism. Whatever the reaction is, it logically stands on a different level of analysis than that pertaining to market reactions.

The Kyoto countries have obviously assumed that other governments will imitate their behavior, expecting that it will be easier to enlist other countries to an energetic policy course if they lead by example. That is why the European Union countries, most of all Germany, agreed to such extensive advances toward capping emissions under the Kyoto Protocol while most of the rest of the world just looked on. The Europeans hope that their strategy will be perceived as fair and will prompt others to follow suit. Whether that will happen remains to be seen. So far no imitating reactions have been observed, apart from some symbolic actions, aimed at pleasing and reassuring the world, in China.

Game theory—a well-established tool in the economic and social sciences, in biology, and in conflict research—analyzes the strategic interactions between decision makers and attempts to forecast their results. From the perspective of game theory, there isn't much to be said in favor of the hope that the advances will bring about the desired results. In fact, rather the opposite could be expected, since if, thanks to the efforts of the Europeans, the world remains cool and the prices of fossil fuels fall, why should other countries commit to painful emissions caps? Since the goal of keeping our planet livable and energy prices low is brought nearer without the developing and emerging countries chipping in, doing nothing politically and letting the markets react by buying the carbon relinquished by the Kyoto countries is the more attractive policy option from a national point of view.

It is true, of course, that negotiating parties aren't always selfish and don't always take hard bargaining stances. As behavioral game theory emphasizes, humans are also guided by the principles of reciprocity and fairness.[15] However, in the case at hand it is entirely unclear what fairness could possibly mean. Everyone wants to be fair, but everyone understands fairness differently. The industrialized countries consider it fair for each country to commit to the same percentage of CO_2 emissions reduction. The developing countries, most of them populous, stick to the position that each country should receive the same per-capita emissions rights, with the hope that such emissions rights can then be sold to the industrialized countries at a profit.[16] The emerging and fast-growing countries of Asia, above all China, maintain the position that they should be granted the right to catch up to the emissions of the industrialized countries since the nineteenth century before they agree to the same reduction goals. Also heard is the argument that it would be fair if the new ownership rights to the air that the certificates represent were used to fight poverty in the world. It will be extraordinarily difficult to bring all these positions to common ground and to achieve an amicable agreement on emissions caps. In the climate negotiations for a new protocol, each delegation's chief will carry fairness in his arsenal as an argument and will elaborate on what that means in his opinion, tenaciously seeking to maximize his own country's advantage. This is one reason why the emissions reductions of the European countries up to now have not impressed the Chinese negotiators at the various climate summits, let alone prompted them to reciprocate.

Economic game theory has already managed to express some aspects of these strategic interactions in formulas, and to state them more precisely as mathematical theorems. If several actors should produce a public good for general use and one of them shoots ahead and sets facts on the ground, the others will take these facts as given and add those they consider sensible. The more advance work one of them has done, the less the others will gain from contributing themselves, and therefore they will opt for contributing less. The one acting first must always make the largest contribution.[17] This is particularly valid when a country commits to "green" technologies and makes clear to its negotiating partners that it could phase out conventional technologies without much difficulty and reduce its CO_2 emissions. This prior contribution will be fully taken advantage of.[18]

It is even likely that the pressing ahead of the well-meaning countries will induce other countries to free-ride to such an extent that, when all countries are taken together, less is accomplished than if an effort had been made to reach a symmetrical negotiation outcome involving every party.[19] If it is clear that either everyone does something or no one does anything and the world falls apart, each negotiator will have a high incentive to support "green" policies. By refusing he would save his own costs, but he couldn't profit from the public good that would have resulted otherwise, namely a reduction of the greenhouse effect and lower fossil-fuel prices in the world market. This would make him think twice before blocking the initiative.

It is different if a single decision-making country can enjoy a free ride without fear that the other countries will back out. Such a country saves its own cost of contributing to the joint abatement effort, but loses only the portion of the common good that would have resulted from its own contribution. The risk is very high that under such conditions no one but the pioneering countries will chip in. And even if everyone makes a contribution under such conditions, it is still possible that the sum of all efforts will be less than if no one had tried to lead by example.

Seen in this light, it is highly advisable for the EU countries and the other Kyoto countries to take a hard negotiating stance in all climate summits to come. It may be true that their pioneering efforts provided the psychologically necessary first impulse to start the whole thing rolling, but if so they shouldn't expand such pioneering efforts. The good will has already been demonstrated. It must be made clear to all other negotiating parties that they may wreck the entire climate-change program if they don't participate. If the Kyoto countries really want to achieve climate protection (including the resulting decrease in the prices of fossil fuels), they must bring a convincing threat strategy to bear. Such a strategy might include imposing import duties on the CO_2 content of imports from opt-out countries, as President Nicolas Sarkozy of France has suggested. It would also be possible to require importers of merchandise to buy CO_2 emissions certificates for the imports they bring in. In the end, only a hard negotiating strategy will be successful.

Source taxes: A supply-side policy

The emphasis on joint demand restraint in the previous sections may be a bit disappointing for the reader, insofar as the focus of this book has been on the supply side. However, this book is driven by an attempt to give a true and responsible assessment of a serious problem that has the potential of jeopardizing the future of mankind, not by an attempt to write a novel.

Still, there is one supply-side policy that is worthy of discussion after all. This policy lies outside the viewpoint of traditional environmental policy, but it follows straightforwardly from the logic laid down in this chapter and the preceding one. If resource owners convert their natural capital into financial assets too quickly, then we should make their natural capital more attractive or, alternatively, make their financial assets less enticing!

To understand this logic, it is useful to reconsider the nature of the market failure described in chapter 4. As has been explained several times, the main reason for market failure in resource extraction is the negative climate externality resulting from the stock of carbon in the atmosphere, which reduces the true rate of return on man-made capital.[20] Present generations can bequeath wealth to later generations in the form of man-made and natural capital, particularly fossil carbon. In order to attain a socially optimal portfolio in view of the global warming problem, the structure of this wealth should be chosen so that the rate of appreciation of the fossil fuels *in situ* equals the rate of return to capital, net of the global-warming damage. As was shown in chapter 4, this is the condition for intertemporal Pareto optimality.

However, as was also shown there, this is not the path that markets choose. Instead, the rate of appreciation of the stocks *in situ* tends to equal the sum of the rate of return to capital and the probability of expropriation. Thus, the appreciation rate of the stocks *in situ* (and hence the rate of price increase of the carbon extracted) is too high, because it compensates for a private expropriation risk that is no social risk and because it balances out an interest rate that exceeds the true social rate of return by the damage from global warming.

A theoretically correct way to compensate for both market failures would be a subsidy on the stock of carbon *in situ* or a tax on the financial capital into which this stock can be converted. Insofar as a subsidy is costly for the world and would make the rich resource owners even richer, only the tax remains as a meaningful option. The tax on financial assets, or, equivalently, on the capital income earned on these assets, would prompt the resource owners to leave a larger part of their wealth below ground, tilting their private wealth portfolio in the direction of a socially optimal portfolio with less man-made capital and more natural capital. The carbon resources would be extracted more slowly, as the extraction and price paths would have to flatten. Today's carbon prices would be higher than without such a tax, and the flow of fossil fuels would initially be lower. However, in the distant future resources would be cheaper and more fossil fuel would be available for consumption. The tax would therefore reduce the market failure, potentially resulting in an intertemporal Pareto improvement.

Admittedly, it wouldn't be easy to tailor the policy measures so that present and future generations would both be made better off in terms of the consumption they can afford. Much will depend on how the source-tax revenue from taxing the petrodollars' returns will be spent—that is, whether it will used for public transfers that finance today's consumption or whether it will be re-invested in infrastructure projects or other types of capital goods. Nevertheless, in principle, the policy could be designed to create a win-win situation for present and future generations.

The problems with such a tax solution are less theoretical than practical. It will not be easy to establish a worldwide system of source taxes on capital income. Capital income taxes already exist in practically all tax systems around the world. However, as a rule, taxes on interest income are levied in accordance with the residence principle, pursuant to the OECD's model double taxation convention of 1977.[21] Interest income, therefore, is typically taxed not where it arises but at the investor's place of residence.

Countries often levy source taxes on interest income in addition to residence-based taxes, but those taxes aren't very important. In many cases taxpayers can deduct them when they declare their interest income at home, which implies that they are irrelevant for capital movements.

The situation is different with capital income from foreign direct investment, i.e., corporate profits belonging to foreign shareholders. As long as corporate profits are retained, the profits are nearly always taxed at the firm's location. The taxes on corporate profits are residence taxes from a legal point of view, but from an economic point of view they are source taxes.

To the extent that the profits are paid out as dividends, they are often taxed both in the firm's and the investor's country of residence. In addition, there are often personal capital-gains taxes on the appreciation of stocks in the country of residence, which are implicit taxes on the retained earnings that create the capital gains.

The residence principle has typically resulted in very light taxation of the capital income of investors from countries with weak tax systems. This principle has also made possible the existence of tax havens that offer investors from all over the world the possibility of collecting their interest income via intermediate companies located there, so that part of the tax burden can be circumvented.

Because it can be assumed that most resource owners are among the investors whose capital income is taxed lightly because of the residence principle, a worldwide switch to source taxes would change the tax situation for them rather dramatically. The returns that had largely escaped taxation would, for the first time, be subject to a substantial tax burden, triggering the desired restructuring of wealth portfolios toward more resource conservation.

It might be feared that resource owners would in this case move their financial investment companies to tax havens, such as the Bahamas, Guernsey, Jersey, or the Cayman Islands. However, this fear is not justified, because the tax havens have no income sources of their own. The capital income channeled through their banks originates from the industrial countries where the petrodollars are ultimately invested, and the source taxes can be levied there. If the industrial countries uniformly applied source taxes on capital income, the tax havens would cease to exist as such.

The move to a source-tax system is less problematic than it sounds, as it would exert no significant influence on the functioning of the capital markets of the industrialized countries, since the capital incomes of the investors residing in such countries are taxed there anyway. Whether these capital incomes are taxed at the source or in the countries of residence doesn't make much difference as long as the tax rates are similar. This is important insofar as no intertemporal distortions in the form of reduced saving and capital formation within the industrialized countries must be feared when the world moves to a system of source taxation of capital incomes.

However, the industrialized countries would have to harmonize their tax systems in order to avert triggering a competition that would see their tax rates engaging in a race to the bottom. Without a harmonization of tax rates, each country would have an incentive to underbid its neighbors to attract more capital at their expense, and in the end the source taxes would be eroded.[22]

This coordination problem can be solved. In fact, the OECD has been surprisingly successful with its most recent attempts to strengthen worldwide taxation at source. Since the OECD announced in 2008 that it would release a blacklist of countries regarded as tax havens, nourishing the expectation that those countries would then face political sanctions by other countries, quite a few of the countries have complied, changing their tax laws and supervisory systems and participating in bilateral agreements to exchange tax information. Switzerland has even decided to loosen its bank-secrecy law. Figure 5.4 demonstrates the remarkable success that this harmonization attempt had. With this experience, the OECD will no doubt be able to harmonize tax systems further.

It is debatable whether source taxes on capital income would be able to move the world's fossil-fuel extraction paths sufficiently in the right

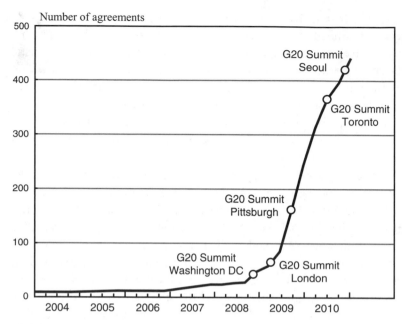

Figure 5.4
Cumulative tax information exchange agreements since 2004. Source: OECD Tax Information Exchange Agreements (http://www.oecd.org/document), January 20, 2011.

direction. Source taxes may have to be established alongside the Super-Kyoto system. Nevertheless, source taxes on capital income offer a solution that is more compatible with a free-market system than a centrally managed system of emissions certificates. If what is sought is a tax solution to internalize the greenhouse gases' external effects, this would be it.

Carbon tax terrors

Trying to slow down global warming with a source tax on capital income sounds far-fetched to a traditional environmentalist. He would be inclined to favor a tax on carbon waste emissions, assuming that this tax would reduce carbon consumption by raising the carbon price. This idea can be traced back to the traditional tax on environmental externalities first proposed by the British economist Arthur Cecil Pigou in 1920.[23] Recently the Stern Review, which triggered the current round of debate on climate change, advocated a carbon tax as the world's main policy tool for sending a uniform price signal around the world.[24]

However, the Pigouvian logic is basically static and doesn't capture the dynamic, intertemporal considerations of resource owners that determine the speed of resource extraction and the global warming resulting from it. Furthermore, the externality that has to be fought depends on the stock of carbon in the atmosphere, whereas the tax advocated is on the flow of carbon emitted. This distinction isn't trivial. It is similar to that between a lake's water level and the water flowing into it from a creek. Though the two values are closely related, they are conceptually different, particularly when filling the lake takes hundreds of years.

A correctly designed Pigouvian tax would have to track the polluters responsible for the stock of anthropogenic carbon in the air and would have to burden their share in this stock with an annual tax whose rate should, in each year, equal the marginal damage that would be caused by a small increment to this stock. Conceptually, we would have to pump the atmospheric carbon into big waste containers labeled with the polluters' names and apply the carbon tax rate to their contents, or equivalently we would have to establish an individual carbon account for each polluter and apply, in each period, a tax on the share in the stock of anthropogenic carbon waste accumulated in the atmosphere reported on this account. Though the time path of the tax rate would have to be exogenously given from the viewpoint of an individual resource owner, as is usual with Pigouvian tax approaches, the tax rate in each period would nevertheless have to equal the marginal damage associated with the then-existing aggregate stock, i.e., the CO_2 concentration in the air that prevails in the period in question. To the best of my knowledge, no one has ever proposed or analyzed such a carbon tax—probably because administering it would be far too demanding, since it would be necessary to trace the tax debtors over hundreds if not thousands of years after they emitted the carbon. Polluting companies come and go, and all too often they end up in bankruptcy, which would wipe out the existing tax debt. It is therefore not possible to establish a true Pigouvian tax on carbon.

What has been analyzed and proposed instead is a proxy tax on carbon emitted. Indeed, if calculated properly, such a tax could mimic a true Pigouvian tax, but the calculation would be prohibitively complicated. The theoretically correct proxy tax rate on the flow of carbon waste would, at each point in time, have to equal the then-prevailing present value of all additional damage from this point until eternity resulting from an additional unit of carbon dioxide being emitted.[25] Apart from the fact that tax laws don't use present-value formulas, the difficulty in the calculation of the tax rate lies in knowing the future time path of damages resulting from an additional unit of carbon extracted today, which itself depends on the entire time path of the stock of carbon that will be extracted from now to infinity.[26]

A major difficulty in the implementation of a tax on the flow of carbon lies in the fact that, because of the stock-flow distinction, its economic effects result from the change of the tax rate over time rather than its sheer size. This can best be understood by considering an *ad valorem* tax on the market value of the flow of carbon consumed, abstracting for a moment from extraction costs. With a constant tax rate, such a tax is basically a cash-flow tax, which effectively makes the government a co-owner of the resource.[27] It generates a revenue without giving the original resource owners the chance to lower the present value of the tax burden by changing the extraction path. It is as if the original resource owners had a silent minority owner in the resource fields who accepts all decisions but wants to have a share of the dividends coming out of them. With a minority owner by their side, the majority owners wouldn't choose another extraction strategy than the one they would choose if they were the sole owners, provided that the dividends are shared in fixed proportions. But things would be different if the minority owner's share of the dividends would rise or fall with the passage of time. When this share fell gradually, which would correspond to a falling *ad valorem* tax rate, the majority owners would have an incentive to defer extraction, because that way they could secure a higher fraction of the dividends for themselves. And when it rose, they would want to extract earlier so as to get their money before the government's demands became larger. This would be the Green Paradox. Only a falling *ad valorem* tax rate would give the desired incentive to slow down extraction.[28]

If a unit tax rate rather than an *ad valorem* tax rate were used, a favorable result could be achieved more easily, as increasing resource prices imply that a constant unit tax translates into an *ad valorem* tax with a declining rate. With a constant unit tax, resource firms have an incentive to postpone extraction because they can reduce the present value of the tax burden or, equivalently, because the silent government co-owner doesn't mind having a declining ownership share.[29]

It may help the reader to see these results in terms of the concept of pricing pressure, defined above as the present value of the (absolute) price wedge ΔP that would be caused by "green" policy measures if the extraction path did not react. Taxation is just one way to create such a price wedge by shifting the aggregate carbon demand curve downward. (See figure 4.4.) With a tax interpretation, ΔP would be equal in size to a unit tax rate on carbon extraction. If ΔP stays constant, its discounted, present value will fall over time, and hence the pricing pressure will diminish, inducing a postponement of extraction. If ΔP stays constant *in present-value terms*, the pricing pressure will be uniform and no supply reactions will occur. If, however, ΔP increases over time, which means that the unit tax rate grows at a rate above the discount rate, extraction will be accelerated and the Green Paradox will prevail.

In the absence of extraction costs, a constant *ad valorem* tax rate would create constant pricing pressure and hence lead to no supply reactions. It marks the borderline case between good and bad reactions of resource owners. If there are stock-dependent extraction costs, as has generally been assumed in this book, the neutrality condition stays the same in terms of the pricing pressure as defined by the present value of ΔP. However, it translates into a subtler formula, with a slightly growing *ad valorem* tax rate, as the co-ownership interpretation of this tax rate would no longer be possible, given that the government doesn't allow extraction costs to be tax-deductible. It can be shown that the borderline growth rate of the *ad valorem* tax rate is now given by the product of the discount rate and the extraction cost-price ratio, which usually is still a small number, close to zero.[30] With a 4 percent interest rate and a 17 percent cost-price share, as in table 5.1, the borderline growth rate of the *ad valorem* tax rate on crude oil consumption, separating the ranges of accelerated and slowed down global warming, would be just two-thirds of a percent per year.

That whether global warming will be slowed down or accelerated will be determined not by the tax rate but by its change over time is a disturbing aspect of the carbon tax, making it a very difficult if not downright dangerous policy instrument. The problems discussed above in the section titled "Paling green" also apply fully to the tax solution. No government is able to credibly promise or guarantee a carbon tax rate that will fall or increase only moderately over time. As the world grows warmer and people get more nervous about climate-caused damages, they will demand stronger government actions. This makes substantial increases in the *ad valorem* tax rate possible, if not likely. Choosing a time path for the tax rate that would tilt the extraction path in the right direction is theoretically conceivable, but it lies beyond the political possibilities of a democratic state that must act upon the growing public unease about climate change.

It is thus little wonder, for example, that the EU Commission called on its member states to introduce a CO_2 tax whose rate is to be increased every year.[31] The Deutsches Institut für Wirtschaftsforschung (German Institute for Economic Research), in its much-commented-on "Greenpeace Study," even recommended raising the tax on carbon by 7 percent above the inflation rate annually.[32] These proposals were formulated without considering the supply reactions of the resource owners, and they are only typical of what would happen around the world if carbon taxes were introduced generally as a policy instrument to fight global warming, as has been recommended by the Stern Review and by many economists. The increase in extraction volumes triggered by such a high growth rate of the carbon tax rate, if applied globally, would, in all likelihood, unleash a strong acceleration of climate change.

One might hope that a unit tax for carbon would offer a better protection against the Green Paradox, as it results in resource conservation when applied with a constant rate. However, over years even a unit tax can easily be raised fast enough to trigger detrimental supply reactions among the resource owners. The danger would be particularly large with a tax adjustment rule that enjoys significant popularity among resource economists and is based on rolling static projections of the currently observable marginal damage from anthropogenic carbon. According to this rule, in each period the unit tax rate must be equal to the current flow of marginal carbon damage divided by the current market rate of interest, because this would be the present value of all future damages caused by a unit of carbon added to the atmosphere in that period that will stay there forever, provided that the marginal damage and the interest rate remain constant. As the marginal damage from global warming increases as a result of the increasing stock of atmospheric carbon, the rule would lead to automatic upward revisions of the carbon tax rate. Nothing would preclude the increase in the unit tax from being sufficiently large to trigger the Green Paradox.[33]

All of this shows that, with a proxy tax on the flow of carbon emissions, the assessment basis is simply too far from the "true" Pigouvian tax on the stock of accumulated waste carbon to be a suitable policy instrument. Trying to steer the economy with a carbon emissions tax is like driving a car the position of whose front wheels doesn't depend on the number of turns of the steering wheel but instead on the speed with which the steering wheel is being turned, and which moreover makes the car turn right when the wheel is spinning left and vice versa. Steering under such circumstances is simply too complicated for real-life decision-making processes. Substantial errors could creep in, making the flow of new carbon dioxide pick up instead of slowing down. Excellent drivers could conceivably manage the trick after successive tries, but normal politicians probably would end up driving the car into a ditch.

Source taxes on capital income, by comparison, are foolproof, since they are already based on a stock, namely the stock of capital that has accumulated parallel to the carbon dioxide stock in the atmosphere. Instead of the change in the tax rate, it is the level of the source tax on capital income that determines what direction the economy takes. Even if the generations of politicians who are responsible for environmental policies keep changing the tax rate in erratic directions, they will always slow down global warming and could never err by unintentionally exacerbating the climate situation.

Apart from these advantages over a tax on carbon flows, the source tax on capital income shares some of the advantages that have been reclaimed for that tax. It gives rise to a unitary price signal that encourages the adoption of reduction strategies because it flattens both the price path and the extraction path and, therefore, increases the current prices of fossil fuels in the world markets. Since this price is valid worldwide, a globally efficient structure of carbon-abatement activities would be ensured. It would make wind turbines, solar heating, nuclear plants, and all the other potentially available strategies for reducing CO_2 more profitable, and it would induce firms to invest time and effort in improving technologies. Furthermore, the source tax would improve the Western countries' tax revenues from the capital income the resource-owning countries earn on their territories.

The only caveat affecting the capital income tax solution is that it may not be strong enough for the fight against global warming. There are two things that limit its impact.

First, the rate of the source tax has to be capped at the level at which normal capital income in the industrial world is taxed. Though a general switch to a source-tax system is possible, there are many practical and legal reasons why resource owners can't be burdened with higher capital income taxes than normal taxpayers. Therefore, the maximum rate of a source tax would be equal to the normal tax on capital income in the Western countries. Increasing this tax rate further would require raising the rates of all taxes on capital income in the industrialized countries, but that might be akin to using one evil to shoo another away. Higher rates for capital income taxes lead to spending on consumption instead of saving and investment. To some extent this might be desirable if the Pareto improvement is to be tailored to the advantage of the present generation. However, excessive tax rates would increase the present generation's consumption to the detriment of later ones, with the result that too little man-made capital will be bequeathed.

Second, levying taxes on capital income might fail to compensate sufficiently for the acceleration of resource extraction resulting from the fear of expropriation. To see why this is so, consider an extreme case and assume that the capital income tax rate is 100 percent, making the net-of-tax interest rate that resource owners could earn in the capital market zero. In this case, the conservation motive for the resource owners would be maximal, and, without an expropriation risk, even the slightest expected price increase would induce them not to sell their resources today. The current price would therefore be astronomical, and environmentalists would be happy. Now suppose there is also an expropriation risk. In this case, resource owners would have a strong incentive to extract their resources quickly to avoid losing their wealth to the members of a rival regime that might take over. They would choose a path with as much extraction in the present and as little in the future as possible, so that the resulting rate of appreciation of the resources *in situ* would be high enough to compensate for the probability of expropriation. But what if the annual probability of expropriation lies above the true economic rate of return on capital, net of the global-warming damage, to which mankind should equate the rate of resource price appreciation for efficiency reasons? In this case extraction would still proceed too quickly. Thus, obviously, even a confiscatory taxation of all capital income may not be enough to enforce the right extraction path. If the maximum source tax the industrialized world could impose on the capital income of resource owners is limited to the normal capital income tax rate, the tax solution may not be sufficiently powerful to compensate for both the global-warming externality and the risk of expropriation, both of which work together toward bringing about excessively fast resource extraction. Thus, in the end, there appears to be an even stronger reason for combining the source tax solution with the Super-Kyoto approach—that is, a worldwide cap-and-trade system.

The two policy tools in fact would complement each other well. As the source tax on capital income in itself already reduces the incentive to overextract, it helps reduce the resource owners' potential resistance against the quantity constraints imposed by a cap-and-trade system. There will be less incentive to circumvent such a system through black-market activities, and less incentive to fight it politically. The source tax acts as a further reinforcement of the barriers mankind should erect against overextraction.

More forests

"Even if I knew that the world would end tomorrow, I would still plant another apple tree today." We should keep these words (attributed to Martin Luther) in mind when it comes to afforestation as a tool for battling global warming, even if the world isn't about to end tomorrow and even if more than just a few apple trees would be needed. Humanity would do well to wield hope and energy to curb global warming.

Clearly we will have to slow down the extraction of fossil fuels if we want to slow down global warming. Nuclear power, wind farms, and perhaps solar power stations located in the world's deserts will help compensate for the resulting energy deficit, but they only generate electricity and, being constrained by their limited storage capacity, they aren't perfect substitutes for hydrocarbons. Transportation on land, on water, and particularly in the air will remain dependent on fossil fuels for a long time. Biofuels will not and should not compensate for them, because of the increasing rivalry with land needed for food production.

It would be ideal if there were a technical tool that would enable us to reduce the output of carbon dioxide without having to abstain from using fossil fuels. But unfortunately, such a simple and straightforward tool doesn't exist. Chemistry offers no way to get around obtaining energy from hydrocarbons without releasing an amount of CO_2 that is strictly proportional to the combusted carbon.

There are only two technically possible ways to decouple the greenhouse effect from energy consumption. One is sequestration of CO_2, discussed in depth in chapter 2. It should be explored further. Still, the enormous volumes of carbon dioxide to be stored and the significant dangers sequestration poses put a damper on such hopes. The other possibility is forests.

As was discussed in chapter 3, wood consists largely of reduced carbon—carbon that has been stripped by photosynthesis of the accompanying oxygen molecule that had turned it into carbon dioxide. The carbon trapped in stems and roots can't do any damage to the atmosphere.

One square kilometer of Brazilian rainforest stores 20,000 tons of carbon (52,000 tons per square mile).[34] In all, the Brazilian forests hold 85 gigatons of carbon. That is more than one-tenth of the stock of carbon floating in the atmosphere, and nearly one-half of the anthropogenic contribution to that stock since industrialization, which is 200 gigatons. (See table 4.2.)

On average, the planet's forests store about 13,000 tons of carbon per square kilometer (34,000 tons per square mile), which, with a total forest cover of 41 million square kilometers (15.8 million square miles), represents 530 gigatons of carbon. This is about 82 percent of the carbon stored in all the biomass on the planet (which amounts to 650 gigatons), and more than 2.5 times the anthropogenic carbon in the air.

If we manage to increase the world's forest cover, we will slow down global warming without having to alter energy consumption. We just take the fossil carbon out of the ground, burn it, and let the sun, through photosynthesis, store it in the trees again. So this is the technical lever that we can pull to forestall climate change. And it is a lever with no ill side effects. On the contrary, forests not only stabilize the world's climate; they also stabilize the local microclimate, offer protection and a habitat for many species, and prevent karst formation. And Italy and Croatia would be much more livable if we could replant the forests that the Romans cut down for shipbuilding.

Unfortunately, this lever is currently being pushed the wrong way. Despite significant afforestation programs in individual countries, forest areas are diminishing instead of increasing around the world. Overexploitation has gained dangerous momentum in recent years not least because of the absurd idea of clearing forests to turn them into arable land for biofuel production. Each year, 129,000 square kilometers (50,000 square miles) of forests are destroyed and only 56,000 square kilometers (22,000 square miles) are replenished through reforestation or afforestation. The net loss of forests per year, according to IPCC estimates, amounts to 73,000 square kilometers (28,000 square miles).[35]

John Houghton, one of the pioneers of climate research, estimates the loss to be 94,000 square kilometers (36,000 square miles).[36] The mean value of these two estimates is 84,000 square kilometers of lost forest per year (32,000 square miles), equivalent to the size of Ireland.

If two-thirds of the forests destroyed yearly end up as carbon dioxide, and the rest as construction timber or soot (thus remaining protected from oxidation), an additional 4.1 gigatons of carbon dioxide are released into the atmosphere, equivalent to 1.1 gigatons of carbon. This is close to the 1.5-gigaton yearly "carbon footprint" of land, sea, and air transportation.[37]

As was discussed in chapter 1, overall annual carbon dioxide emissions from combustion of fossil carbon and production of cement amount to 7.4 gigatons of carbon. The 1.1 gigatons of carbon that are released through forest destruction thus account for about 13 percent of the annual man-made carbon dioxide emissions. This is, of course, an extremely cautious estimate that is based solely on the wood. Often, however, the clearing of forests destroys ground-level flora as well, releasing a large additional amount of CO_2. When boggy soils and swamps are drained, greenhouse gases (particularly methane, one of the most powerful greenhouse gases) are released as organisms decay. In special cases, these emissions can be several times the direct CO_2 emissions from burning the wood. Some of the carbon stored in construction timber will also be released eventually, when the buildings they form a part of are demolished and the timber is burned or left to decay. For purposes of long-range CO_2 calculation, nearly half will have to be added to the cited CO_2 emissions due to the destruction of forests.

If deforestation were stopped and instead an area the size of Ireland were afforested every year, man-made CO_2 emissions would decrease by about 70 gigatons of carbon by the middle of the century. This represents about 8 percent of the carbon that, according to the business-as-usual scenario of the Stern Review, will be emitted until then. Relative to the pitiful reduction goals committed to under the Kyoto Protocol, that is indeed quite a respectable amount. Maybe we should follow Martin Luther's advice.

Instruments and goals

We humans (and all other living organisms on Earth) consist of carbon compounds. We are built out of carbon, and we derive our energy from it. We must, in order to live, eat other organisms, since we can't capture carbon from the atmosphere as plants do. We are complements of the plants, burning carbon with the help of the oxygen that plants emit as waste, and playing a part in the carbon cycle. For a time, our minds can unfold, and we can communicate with other people. Then the structure comes apart once again. Fortunately, knowledge, preferences, and opinions can be passed on to future generations.

In the fight for survival between living beings a balance has been established in which the spread of a species is determined by the availability of food and materials for building its dwellings. Mankind has perceived this as a Malthusian trap. The number of people who can live is limited by the amounts of food, air, and room available for them.

Our ancestors, about eight generations ago, escaped the Malthusian trap by finding ways to live off fossil carbon. Fossil carbon made it possible to convert the land producing the fodder needed for draft and pack animals into food, and it provided the extra energy that turned the wheels of the world's economy with ever-increasing speed, allowing more and more humans to make a living on the limited surface of this planet.

Not only did fossil fuels bring us bread; they also gave us the energy that has powered our civilization. But they have made us dependent. Even worse, they have produced waste that we just don't know what to do with. We observe with concern that the carbon stocks below ground are steadily decreasing, while those in the atmosphere increase. Rising energy prices make us fear for our prosperity, and rising temperatures make us fear for our well-being.

It will not be easy to live without fossil carbon, because there are now so many of us—ten times as many as were living when industrialization began. We, who in no small measure owe our very existence to the use of fossil fuels, will have a hard time doing without them. With a ballooning population, an excess of carbon in the atmosphere, and imperfect and limited technologies for replacing fossil fuels, mankind could get stuck once again in the Malthusian trap.

Nature is already visibly affected by global warming, and even as the grim predictions of the climatologists give cause for concern, now come the Chinese, the Indians, and many others who rightly also want to live better, adding huge amounts of CO_2 with their blistering economic growth. The situation will become critical because 100 percent of the world's population wants to live like the 15 percent who populate the developed countries.

Although (or, rather, because) the planet is warming, people in northern latitudes will be able to live well on their expanding fertile territories. But when Siberia and Alaska acquire more temperate climes, will we accept sacrificing Africa and India in exchange? Migration from the old to the new living areas will not occur without tensions, crises, and much violence.

There will also be strife triggered by the increasing scarcity of resources. The West has staked claims in the Middle East. China is trying to increase its influence in Africa. The United States has extended a hand to Kazakhstan, home to large gas and oil deposits. The attempt to extend NATO to Russia's southern flank raised tensions that discharged themselves explosively in Georgia. There will be other such flash points, because wrangling for resources will increase as resources become scarcer.

Even now, the political uncertainty gives rise to insecurity and fear among the owners of fossil-fuel resources, prompting them to extract their resources quickly. The ruthless exploitation of the past decades is not a result of the market economy. Among the hallmarks of the market economy are secure property rights, on the basis of which trade can occur. If property rights are not secure, there is no market economy; there is anarchy and chaos. If the owners fear the loss of their natural resources through war or insurrection, they will prefer the safety of a Swiss bank account, but such an account can be filled only if they sell their resources as rapidly as possible and risk the collateral damage represented by climate change. These interrelationships should be made clear to those wishing to "threaten" the dictatorships of this world with Western-style democracy. There are many reasons to fight dictatorships, but climate problems aren't among them. Maintaining peace on Earth is the basic prerequisite for mankind to husband its resources sensibly. In this respect the West should cast a critical glance upon itself and acknowledge how much it may be at fault.

Quantity management within the framework of a Super-Kyoto system with emissions trading, uniting all consumer countries in a seamless global cartel, is a desperate but indispensable emergency answer to predatory extraction, even though it embodies a rationing system that evokes bad memories among Europeans who experienced the postwar years. Such a system should be introduced very swiftly, since a piecemeal, incremental approach would reinforce the resource owners' fears and would accelerate what it aims to slow down. Whether expropriation by rivals, by a superpower, or by the UN is expected, the allure of a Swiss bank account stays the same. The Super-Kyoto system should be accompanied by a worldwide move to source taxation of capital income in order to spoil the resource owners' appetite for financial assets and induce them to keep a larger share of their wealth underground.

Idiosyncratic national measures to curb the demand for carbon are mostly unproductive, unable to help mitigate global warming, because demand policies can't succeed if they are carried out only by a subset of countries. The demand restraint merely reallocates the extracted fossil fuel to other countries; what is more, it may create pricing pressure that, if expected to grow over time, will backfire by inducing the resource owners to anticipate the destruction of their markets and sell their resources prematurely. What the oil sheiks and coal barons have for decades perceived as "green" saber-rattling has contributed to the selling off at reduced prices of carbon resources that has caused the world's economy to boom at the expense of the environment we will bequeath to future generations. It is time now to stop striking poses and finally engage in battle with global warming.

Some would like to battle global warming by resorting to the natural carbon sources with which our ancestors contented themselves. They want to fuel their modern comforts with bioenergy. But they forget that these natural sources had already proved insufficient before the use of fossil fuels liberated land for food production and allowed the world's population to swell. With all the mouths that have to be fed today, it is simply not possible to convert our farms into biofuel refineries. In 2008, the attempt to turn back the wheel of history provoked hunger riots in 37 countries—and not only among those fond of tortillas.

European politicians in particular are betting on "green" energy and are keen on playing the exemplary boy. They gleefully set high CO_2-reduction goals and boast of making great strides toward turning their countries "green." The goals that the politicians pursue have to be honored and can be acknowledged from the socioeconomic-welfare perspective developed in chapter 4 of this book. Bequeathing future generations more carbon below the ground and less man-made capital above the ground makes it possible to make future generations better off without harming present ones, and could even be used to shift the benefit to present generations without harming future ones. However, goals have to be distinguished from instruments. The road to hell, after all, is paved with good intentions.

When attempting to design "green" policies, politicians and the general public tend to concentrate solely on the laws of physics, turning a blind eye to those of economics. But if they really want to tackle global warming effectively in an interconnected world economy, it is precisely the laws of economics that they must heed to in order to exert the necessary influence on the energy markets, resource suppliers, and consumers. All too often they appear naive, if not overly ideological, being more interested in doing something than in the results of their actions. How else could they believe that they will be able to reduce carbon emissions solely by curbing demand in only some of the world's countries? How else could they disregard the fact that the way the climate change is ultimately determined by supply, which is in the hands of oil sheikhs, Putin's oligarchs, and coal barons? How else could they fail to see that by subsidizing wind and solar power by way of feed-in tariffs they will be unable to curb European carbon dioxide emissions, because these are already determined by the European cap-and-trade system? How could they overlook the terrors that the development of bioenergy will bring to this world, and willingly partake of the guilt by fueling with their "green" policy measures what only can end in even more riots, conflicts, and strife?

Please don't get me wrong. Mankind has to fight climate change. An ever-increasing carbon concentration is dangerous, and countervailing actions are urgently needed. But the policy tools chosen have to be effective. It isn't enough to make charitable donations, and then bask in a warm glow, if the gifts never reach the addressees. No progress will be made if economic resources are wasted in a jungle of mutually contradictory policies that make no contribution to slowing down global warming and will, in the end, just frustrate the voters and induce them to shy away from spending any more money on ineffective "green" policies. For the sake of our descendants, policy makers and society have to steer away from the ideologies and the wishful thinking that have already spawned a host of harmful decisions, and instead pursue pragmatic and efficient policies that will actually do something to save the only planet we have.

Notes

Preface

1. H.-W. Sinn, *Das grüne Paradoxon. Plädoyer für eine illusionsfreie Klimapolitik* (Econ), first edition 2008, second edition 2009. For more references, see chapters 4 and 5.

2. I refer here to Ottmar Edenhofer and his team at the Potsdam-Institut für Klimafolgenforschung (Potsdam Institute for Climate Impact Research). Mr. Edenhofer is co-chair of Working Group II of the IPCC.

Chapter 1

1. Avogadro's law states that equal volumes of ideal gases contain, at the same pressure and temperature, the same number of molecules. This means that there is the same number of molecules of each gas per cubic meter.

2. See J. T. Houghton, *Global Warming* (Cambridge University Press, 2004), p. 16; B. Klose, *Meteorologie* (Springer, 2008), p. 116; W. Weischet and W. Endlicher, *Einführung in die allgemeine Klimatologie* (Gebrüder Bornträger, 2008), p. 160.

3. See Houghton, *Global Warming*.

4. See N. Stern, *The Economics of Climate Change: The Stern Review* (Cambridge University Press, 2007), p. iv; S. Solomon, D. Qin, M. Manning, Z. Chen, M. Marquis, K. B. Averyt, M. Tignor, and H. L. Miller, eds., *Climate Change 2007: The Physical Science Basis* (Cambridge University Press, 2007), p. 435.

5. B. Frenzel, B. Pecsi, and A. A. Velichko, eds., *Atlas of Palaeoclimates and Palaeoenvironments of the Northern Hemisphere* (INQUA/Hungarian Academy of Sciences and Gustav Fischer, 1992).

6. Ibid.

7. This phenomenon was described first by Joseph Fourier in 1824. See J. Fourier, "Remarques générales sur les températures du globe terrestre et des espaces planétaires," *Annales de Chemie et de Physique* 27 (1824): 136–167; J. Fourier,

"Memoire sur les températures du globe terrestre et des espaces planétaires," *Memoires de l'Academie Royale des Sciences* 7 (1827): 569–604. Further milestones in research were set by J. Tyndall ("On the absorption and radiation of heat by gases and vapours, and on the physical connection of radiation, absorption, and conduction," *Philosophical Magazine* 22 (1861): 169–194, 273–285) and S. Arrhenius ("On the influence of carbonic acid in the air upon the temperature of the ground," *Philosophical Magazine* 41 (1896): 237–276). Svante Arrhenius estimated that a doubling of the CO_2 content in the atmosphere would lead to a 5–6°C increase in temperature, a value remarkably close to present estimates. Roger Revelle and Hans Suess were the first to call attention to the danger posed by increasing CO_2 emissions to our climate ("Carbon dioxide exchange between atmosphere and ocean and the question of an increase of atmospheric CO_2 during the past decades," *Tellus* 9 (1957): 18–27).

8. The source of the data mentioned here is R. K. Pachauri and A. Reisinger, eds., *Climate Change 2007: Synthesis Report* (IPCC, 2007).

9. See Houghton, *Global Warming*, p. 16; J. Jucundus, "Zusammenhänge und Wechselwirkungen im Klimasystem," in *Der Klimawandel—Einblicke, Rückblicke und Ausblicke*, ed. W. Endlicher and F.-W. Gerstengarbe (G&S Druck und Medien, 2007).

10. J. T. Kiehl and K. E. Trenberth, "Earth's annual global mean-energy budget," *Bulletin of the American Meteorological Society* 78 (1997): 197–208.

11. One exception is the water vapor an airplane leaves in its wake in the form of condensation trails (contrails) in the upper atmosphere. It may need up to 100 years to precipitate as rain. But it does eventually disappear completely. See W. Zittel and M. Altmann, "Birgt eine Wasserstoffenergiewirtschaft höhere Klimarisiken als die Verbrennung fossiler Energieträger?" *Energie* 45 (1994): 25–29.

12. D. Archer, "Fate of fossil fuel CO_2 in geologic time," *Journal of Geophysical Research* 110 (2005): 5–11; D. Archer and V. Brovkin, "Millennial atmospheric lifetime of anthropogenic CO_2," 2006, submitted to *Climate Change* (http://www.pik-potsdam.de); G. Hoos, R. Voss, K. Hasselmann, E. Meier-Reimer, and F. Joos, "A nonlinear impulse response model of the coupled carbon cycle-climate system (NICCS)," *Climate Dynamics* 18 (2001): 189–202.

13. Solomon et al., *Climate Change 2007: The Physical Science Basis*, p. 212. See J. T. Houghton, Y. Ding, D. J. Griggs, M. Noguer, P. J. van der Linden, X. Dai, K. Maskell, and C. A. Johnson, eds., *Climate Change 2001: The Scientific Basis* (Cambridge University Press, 2001), p. 554.

14. Further details can be found in C. D. Schönwiese, *Klimatologie*, second edition (Ulmer, 2003).

15. CSIRO Atmospheric Research, *Key Greenhouse and Ozone Depleting Gases* (http://www.cmar.csiro.au/research/capegrim_graphs.html).

16. L. K. Gohar and K. P. Shine, "Equivalent CO_2 and its use in understanding the climate effects of increased greenhouse gas concentrations," *Weather* 62 (2007): 307–311. It is worth mentioning that the individual greenhouse-

equivalent effects can't be added together, because gases interact in a complex manner and even filter overlapping wavelengths, so that the overall effect can only be calculated using complex climate models.

17. See D. M. Etheridge, L. P. Steele, R. L. Langenfelds, R. J. Francey, J.-M. Barnola, and V. I. Morgan, "Historical CO_2 records from the Law Dome DE08, DE08-2, and DSS ice cores," in *Trends: A Compendium of Data on Global Change* (Carbon Dioxide Information Analysis Center, Oak Ridge National Laboratory, 1998); D. M. Etheridge, L. P. Steele, R. L. Langenfelds, R. J. Francey, J.-M. Barnola, and V. I. Morgan, "Natural and anthropogenic changes in atmospheric CO_2 over the last 1000 years from air in Antarctic ice and firn," *Journal of Geophysical Research* 101 (1996): 4115–4128; D. M. Etheridge and C. W. Wookey, "Ice core drilling at a high accumulation area of Law Dome, Antarctica, 1987," in *Ice Core Drilling: Proceedings of the Third International Workshop on Ice Core Drilling Technology*, ed. C. Rado and D. Beaudoing (Grenoble, 1989); V. I. Morgan, C. W. Wookey, J. Li, T. D. van Ommen, W. Skinner, and M. F. Fitzpatrick, "Site information and initial results from deep ice drilling on Law Dome," *Journal of Glaciology* 43 (1997): 3–10.

18. See G. Marland, B. Andres, and T. Boden, *Global CO_2 Emissions from Fossil-Fuel Burning, Cement Manufacture, and Gas Flaring: 1751–2005* (Carbon Dioxide Information Analysis Center, Oak Ridge National Laboratory, 2008); E. Worrell, L. Price, N. Martin, C. Hendricks, and L. Ozawa Meida, "Carbon dioxide emissions from the global cement industry," *Annual Reviews of Energy and the Environment* 26 (2001): 203–229.

19. International Energy Agency, *CO_2 Emissions from Fuel Combustion*, release 01, CO_2 *Sectoral Approach* (International Energy Agency, 2007).

20. Houghton, *Global Warming*, p. 65, figure 4.3.

21. Solomon et al., *Climate Change 2007*, p. 763.

22. P. D. Jones, D. E. Parker, T. J. Osborn, and K. R. Briffa, "Global and hemispheric temperature anomalies—Land and marine instrumental records," in *Trends: A Compendium of Data on Global Change* (Carbon Dioxide Information Analysis Center, Oak Ridge National Laboratory, 2006).

23. B. L. Otto-Bliesner, E. C. Brady, G. Clauzet, R. Tomas, S. Levis, and Z. Kothavala, "Last glacial maximum and holocene climate in CCSM3," *Journal of Climate* 19 (2006): 2526–2544.

24. See S. McIntyre and R. McKitrick, "Corrections to the Mann et al. (1998) proxy data and Northern Hemispheric average temperature series," *Energy & Environment* 14 (2003): 751–771.

25. Long-Term Meteorological Station Potsdam Telegrafenberg (http://saekular .pik-potsdam.de/2007_de/).

26. See Schönwiese, *Klimatologie*.

27. See N. Scafetta and B. J. West, "Phenomenological reconstructions of the solar signature in the Northern Hemisphere surface temperature records since 1600," *Journal of Geophysical Research* 112 (2007), document 24S03.

28. R. E. Benestad and G. A. Schmidt, "Solar trends and global warming," *Journal of Geophysical Research* 114 (2009), document D14101.

29. See M. Lockwood and C. Fröhlich, "Recent oppositely directed trends in solar climate forcings and the global mean surface air temperature," *Proceedings of the Royal Society A* 463 (2007): 2447–2460.

30. M. E. Mann, R. S. Bradley, and M. K. Hughes, "Global-scale temperature patterns and climate forcing over the past six centuries," *Nature* 392 (1998): 779–787; M. E. Mann, R. S. Bradley, and M. K. Hughes, "Northern Hemisphere temperatures during the past millennium: Inferences, uncertainties, and limitations," *Geophysical Research Letters* 26 (1999): 759–762; Houghton et al., *Climate Change 2001*, p. 134, figure 2.20.

31. McIntyre and McKitrick, "Corrections to the Mann et al. (1998) proxy data and northern hemispheric average temperature series."

32. S. McIntyre and R. McKitrick, "Hockey sticks, principal components and spurious significance," *Geophysical Research Letters* 32 (2005), document L03710.

33. National Research Council of the National Academies, *Surface Temperature Reconstructions for the last 2,000 years* (National Academies Press, 2006).

34. See M. E. Mann, Z. Zhang, M. K. Hughes, R. S. Bradley, S. K. Miller, S. Rutherford, and F. Ni., "Proxy-based reconstructions of hemispheric and global surface temperature variations over the past two millennia," *Proceedings of the National Academy of Sciences* 105 (2008): 13252–13257. This is criticized in S. McIntyre and R. McKitrick, "Proxy inconsistency and other problems in millennial paleoclimate reconstructions," *Proceedings of the National Academy of Sciences* 106 (2009), document E10. Another reply is given in M. E. Mann, R. S. Bradley, and M. K. Hughes, "Reply to McIntyre and McKitrick: Proxy-based temperature reconstructions are robust," *Proceedings of the National Academy of Sciences* 106 (2009), document E11.

35. D. Lüthi, M. Le Floch, B. Bereiter, T. Blunier, J.-M. Barnola, U. Siegenthaler, D. Raynaud, J. Jouzel, H. Fischer, K. Kawamura, and T.F. Stocker, "High-resolution carbon dioxide concentration record 650,000–800,000 years before present," *Nature* 453 (2008): 379–382. See also U. Siegenthaler, T. F. Stocker, E. Monnin, D. Lüthi, J. Schwander, B. Stauffer, D. Raynaud, J.-M. Barnola, H. Fischer, V. Masson-Delmotte, and J. Jouzel, "Stable carbon cycle-climate relationship during the late Pleistocene," *Science* 310 (2005): 1313–1317.

36. In the Vostok ice core from the Russian Antarctica station, the 300 ppm found for the Eemian Interglacial 125,000 years ago don't come close to today's values. See Siegenthaler et al., "Stable carbon cycle-climate relationship."

37. See J. E. Harries and D. Commelynck, "The Geostationary Earth Radiation Budget Experiment on MSG-q and its potential applications," *Advances in Space Research* 24 (1999): 915–919; M. Goody and Y. L. Yung, *Atmospheric Radiation—Theoretical Basis* (Oxford University Press, 1995); J. T. Houghton, "Global warming," *Reports on Progress in Physics* 68 (2005): 1343–1403.

38. J. E. Harries, H. E. Brindley, P. J. Sagoo, and R. J. Bantges, "Increases in greenhouse forcing inferred from the outgoing longwave radiation spectra of the Earth in 1970 and 1997," *Nature* 410 (2001): 355–357.

39. See E. Steig, "The lag between temperature and CO_2 (Gore's got it right)," at http://www.realclimate.org. See also J. Severinghaus, "What does the lag of CO_2 behind temperature in ice cores tell us about global warming?" at http://www.realclimate.org.

40. See Steig, "The lag between temperature and CO_2"; J. Severinghaus, "What does the lag of CO_2 behind temperature in ice cores tell us?"

41. "Klimaschutz—Skeptiker fragen, Wissenschaftler antworten," in "Häufig vorgebrachte Argumente gegen den anthropogenen Klimawandel" (http://www.umweltbundesamt.de).

42. Source: Glaciology Commission of the Bavarian Academy of Science (http://www.Glaziologie.de).

43. Various research findings suggest that the ice in West Antarctica is diminishing and that in East Antarctica is increasing. See S. Solomon et al., *Climate Change 2007*, p. 364f. See also M. L. Parry, O. F. Canziani, J. P. Palutikof, P. J. van der Linden, and C. E. Hanson, eds., *Climate Change 2007: Impacts, Adaptation and Vulnerability* (Cambridge University Press, 2007), p. 663.

44. See J. M. Gregory, P. Huybrechts, and S. C. B. Raper, "Threatened loss of the Greenland ice-sheet," *Nature* 428 (2004): 616; D. Dahl-Jensen, J. Bamber, C. E. Bøggild, E. Buch, J. H. Christensen, K. Dethloff, M. Fahnestock, S. Marshall, M. Rosing, K. Steffen, R. Thomas, M. Truffer, M. van den Broeke, and C. J. van der Veen, *The Greenland Ice Sheet in a Changing Climate: Snow, Water, Ice and Permafrost in the Arctic (SWIPA)* (Arctic Monitoring and Assessment Programme, 2009).

45. See J. A. Church and N. J. White, "A 20th century acceleration in global sea-level rise," *Geophysical Research Letters* 33 (2006), L01602, doi:10.1029/2005GL024826; Solomon et al., *Climate Change 2007: The Physical Science Basis*, p. 387.

46. See J. T. Houghton, *The Physics of Atmospheres*, third edition (Cambridge University Press, 2002).

47. T. M. Lenton, H. Held, E. Kriegler, J. W. Hall, W. Lucht, W. Rahmstorf, and H. J. Schellnhuber, "Tipping elements in the Earth's climate system," *Proceedings of the National Academy of Sciences* 105 (2008): 1786–1793.

48. Stern, *The Economics of Climate Change*.

49. Ibid., p. III (Summary and Conclusions). The assertion that a doubling of carbon dioxide concentration can be expected until the middle of the century can be found on p. 15.

50. See table 4.2 in chapter 4 of the present volume.

51. N. Nakicenovic and R. Swart, eds., *IPCC Special Report on Emissions Scenarios* (Cambridge University Press, 2000).

52. "IEA Executive Director: The Climate Challenge is Immense but We Have the Clean Technology," IEA press release, November 12, 2008 (http://www.iea .org); International Energy Agency, *World Energy Outlook 2008*, p. 381f.

53. N. Stern, "Key elements of a global deal on climate change," London School of Economics, April 30, 2008 (http://www.lse.ac.uk).

54. Munich's average temperature is 7.6°C; Milan's is 13.1°C.

55. See, e.g., Parry et al., *Climate Change 2007: Impacts, Adaption and Vulnerability*, chapters 9, 11, 12, and 14.

56. Münchener Rückversicherung, *Annual Review: Natural Catastrophes 2003*, 2004.

57. See Wissenschaftlicher Beirat der Bundesregierung Globale Umweltveränderungen, *Die Zukunft der Meere—zu warm, zu hoch, zu sauer*, Sondergutachten, Berlin, 2006 (http://www.wbgu.de/wbgu_sn2006.pdf).

58. See *Vulnerability and Adaptation to Climate Change in Europe*, Technical Report 7/2005, European Environment Agency.

59. See, e.g., Parry et al., *Climate Change 2007: Impacts, Adaption and Vulnerability*, chapters 6 and 8.

60. See ibid., chapters 10 and 14.

61. See ibid., chapter 7.

62. See H. Welzer, *Klimakriege. Wofür im 21. Jahrhundert getötet wird* (S. Fischer, 2008).

63. See, e.g., Parry et al., *Climate Change 2007: Impacts, Adaption and Vulnerability*, chapter 8.

64. M. Latif, C. Böning, J. Willebrand, A. Biastoch, J. Dengg, B. Schneider, and U. Schweckendiek, "Is the thermohaline circulation changing?" *Journal of Climate* 19 (2006): 4631–4637.

65. H. Sterr, "Folgen des Klimawandels für Ozeane und Küsten," in W. Endlicher and F.-W. Gerstengarbe, eds., *Der Klimawandel—Einblicke, Rückblicke und Ausblicke* (Potsdam Institute for Climate Impact Research, G&S Druck und Medien, 2007), p. 89.

66. See Stern, *The Economics of Climate Change*, chapter 6, p. 143.

67. Ibid., chapter 9, p. 211. The Stern estimate for the mitigation cost agrees roughly with an assertion by the International Energy Agency that measures to halve the temperature increase by the year 2050 would cost a lump-sum amount of $45 trillion, as this would correspond to an annual 1 percent of the world's GDP generated up to that date. See C. Neidhart, "Billionen-Revolution—Die Internationale Energie-Agentur schlägt Alarm," *Süddeutsche Zeitung*, June 7/8, 2008, cover story; International Energy Agency, *Energy Technology Perspectives 2008—Scenarios and Strategies to 2050*, 2008, pp. 224 and 241. The estimated costs vary strongly with the assumptions about technological progress, as future innovations in the energy sector play a central role for the costs of stabilizing carbon dioxide content in the atmosphere; O. Edenhofer, C. Carraro, J. Köhler,

and M. Grubb, eds., "Endogenous technological change and the economics of atmospheric stabilisation," *The Energy Journal*, special issue, 2006 (http://www .pik-potsdam.de).

67. Stern, *The Economics of Climate Change*, p. 185.

Chapter 2

1. G8 Information Centre, *G8 Summits, Hokkaido Official Documents, Environment and Climate Change*, Hokkaido Toyako Summit, July 8, 2008.

2. UNSW Climate Change Research Centre, *The Copenhagen Diagnosis, Updating the World on the Latest Climate Science*, 2009, pp. 50–51.

3. United Nations, *UN Climate Change Conference in Cancun Delivers Balanced Package of Decisions, Restores Faith in Multilateral Process*, press release, Cancun, December 11, 2010.

4. R. Revelle and H. Suess, "Carbon dioxide exchange between atmosphere and ocean and the question of an increase of atmospheric CO_2 during the past decades," *Tellus* 9 (1957): 18–27.

5. Data on CO_2 emissions for the year 2005 from International Energy Agency, *CO_2-Emissions from Fuel Combustion*, release 01, CO_2 Sectoral Approach, 2007 (http://www.sourceoecd.org).

6. European Environment Agency, *EU Greenhouse Gas Emissions Decrease in 2005*, press release, June 15, 2007 (http://www.eea.europa.eu).

7. European Commission, *Limiting Analysis of Options to Move beyond 20% Greenhouse Gas Emission Reductions and Assessing the Risk of Carbon Leakage*, Communication from the Council, The European Parliament, The European Economic and Social Committee and the Committee of the Regions, COM (2010) 265 final, Brussels 2010.

8. G8, *G8 Leader's Summit. Chair's Summary*, L'Aquila, July 10, 2009 (http:// www.g8italia2009.it). See also European Commission, *Statement of President Barroso on the Result of the Meeting of the Major Economies Forum on Climate Change: 2°C Target is now Written in Stone*, press release, July 9, 2009 (http:// europa.eu).

9. Denmark has committed to a reduction equal in size to that of Germany. Luxembourg committed to an even larger reduction (28%).

10. International Energy Agency, *World Energy Outlook 2007: China and India Insights*, 2007.

11. International Energy Agency, *World Energy Outlook 2009, Tables for Reference Scenarios Projections, Reference Scenario World*, 2009; Ifo Institute calculations.

12. International Energy Agency, *Energy Balances of OECD Countries*, 2009 edition, 2009.

13. Cf. International Energy Agency, *World Energy Outlook 2009*.

14. International Energy Agency, *Electricity Information, 2009 IEA Statistics*, 2009.

15. International Energy Agency, *Energy Balances of OECD Countries*, 2009.

16. Renewable Energy Policy Network for the 21st Century (REN21), *Renewables Global Status Report 2009 Update*, 2009.

17. Arbeitsgemeinschaft Energiebilanzen, *Energiebilanzen für die Bundesrepublik Deutschland 2010*, 2010; Arbeitsgruppe Erneuerbare Energien-Statistik (AGEE-Stat), *Entwicklung der erneuerbaren Energien in Deutschland im Jahr 2009*, 2010.

18. International Energy Agency, *Energy Balances of OECD Countries*, 2009.

19. O. Hahn and F. Strassmann, "Über den Nachweis und das Verhalten der bei der Bestrahlung des Urans mittels Neutronen entstehenden Erdalkalimetalle," *Die Naturwissenschaften* 27 (1939): 11–15; O. Hahn and F. Strassmann, "Nachweis der Entstehung aktiver Bariumisotope aus Uran und Thorium durch Neutronenbestrahlung; Nachweis weiterer aktiver Bruchstücke bei der Uranspaltung," *Die Naturwissenschaften* 27 (1939): 89–95; L. Meitner and O. R. Frisch, "Disintegration of uranium by neutrons: A new type of nuclear reaction," *Nature* 143 (1939): 239.

20. See International Atomic Energy Agency, WHO, and UNDP, *Chernobyl, The True Scale of the Accident*, press release, London, Vienna, Washington and Toronto, September 5, 2005 (http://www.iaea.org).

21. See World Health Organization, *Health Report of the UN Chernobyl Forum, Expert Effects of the Chernobyl Accident and Special Health Care Programmes Group «Health»*, 2006, chapter 7, pp. 98–107. The report points out that the estimate of a further 5,000 casualties is uncertain, as it assumes an average dose of radiation that differs little from natural background radiation.

22. See J. Seidel, *Kernenergie—Fragen und Antworten* (Econ, 1990), p. 82.

23. See International Atomic Energy Agency, "Radiation release at Fukushima will not increase much," press release, April 19, 2011; Euronews, "Rosatom's Situation Center Representative Sergey Novikov interviewed by Euronews TV channel" (http://www.rosatom.ru).

24. See C. Clauser, "Geothermal energy," in *Energy Technologies*, Subvolume C: *Renewable Energy*, ed. K. Heinloth (Springer, 2006), p. 486f.

25. See J. A. Plant and A. D. Saunders, "The radioactive Earth," *Radiation Protection Dosimetry* 68 (1996): 25–36. According to the authors, this is a minimum value, because only the long-lived isotopes of uranium, thorium, and potassium were counted, while other short-lived, and therefore particularly radioactive materials that don't occur anymore were left out.

26. L. A. Loeb and R. J. Monnat Jr., "DNA polymerases and human disease," *Nature Review Genetics* 9 (2008): 594–604.

27. T. Herrmann, M. Baumann, and W. Dörr, *Klinische Strahlenbiologie—kurz und bündig*, fourth edition (Elsevier, 2006), p. 16.

28. G. Buttermann, *Radioaktivität und Strahlung* (Verlag R. S. Schulz, 1987), p. 23.

29. Plutonium's half-life is 24,390 years. This implies a mean lifetime of 35,187 years. Half-lives are used only for those substances that decay at a fixed rate. This is not so in the case of carbon dioxide, as about one-fourth of the original volume remains in the atmosphere for an extremely long time. It is possible to convert a half-life into mean lifetime, however. It takes seven half-lives for the radioactivity to decrease to less than 1 percent of the radioactivity of the original material.

30. B. Metz, O. Davidson, H. C. de Coninck, M. Loos, and L. A. Meyer, eds., *IPCC Special Report on Carbon Dioxide Capture and Storage* (Cambridge University Press, 2005), p. 220 ff.

31. See table 4.2.

32. C. Ploetz, *Sequestrierung von CO_2: Technologien, Potenziale, Kosten und Umweltauswirkungen*, Wissenschaftlicher Beirat der Bundesregierung Globale Umweltveränderungen, Externe Expertise für das WBGU-Hauptgutachten 2003 *Welt im Wandel: Energiewende zur Nachhaltigkeit*, (Springer, 2003).

33. The current proportion of oxygen in the atmosphere is 20.946%. When carbon is burned, each carbon molecule acquires one oxygen molecule (which is to say, each carbon atom binds with two oxygen atoms). Carbon has a molar mass of 12.0107 grams per mol, so that the total carbon stock (6,500 gigatons; see footnote to figure 4.7) amounts to approximately 541.19×10^{15} mols (1 mol contains 6.022×10^{23} molecules). Thus, a total of 1082.38×10^{15} mols of oxygen atoms would be removed from the atmosphere if the entire carbon stocks on Earth were burned and no carbon dioxide sequestration be performed. Of this total, over time about 30% of the oxygen atoms bound in carbon dioxide would be detached by photosynthesis and released back into the atmosphere. That means that 70% of the oxygen atoms bound to carbon by combustion will remain as carbon dioxide in the atmosphere or in the oceans. Taking the accepted mass of the atmosphere in kilograms and the average molar mass of the air (28.9644 grams per mol), the oxygen content of the atmosphere would decrease from 20.946% to 20.023%. If all the CO_2 emitted were sequestered, thus preventing photosynthesis from acting on it, all the oxygen would remain bound to carbon. In this case, the proportion of oxygen removed from the atmosphere would be 1.319%, causing the share of oxygen in the atmosphere to decrease to 19.627%. As regards the 30% oxygen detachment through photosynthesis, see J. G. Canadell, C. Le Quéré, M. R. Raupach, C. B. Field, E. T. Buitenhuis, P. Ciais, T. J., Conway, N. P. Gillett, R. A. Houghton, and G. Marland, "Contributions to accelerating atmospheric CO_2 growth from economic activity, carbon intensity, and efficiency of natural sinks," *Proceedings of the National Academy of Sciences* 104 (2007): 18866–18870.

34. With a yearly utilization time of 7,500 hours, the nuclear power plant would generate 9.19 terawatt-hours of electricity annually (1,225 megawatts × 7,500 hours). With an efficiency factor of 38%, this would translate into an anthracite

input of 24.178 terawatt-hours, equivalent to 2.97 million anthracite units, which again translates into approximately 2.9 million tons of anthracite, considering a calorific value of 1.024 anthracite units per kilogram. With a specific weight of 1.4 kilograms per liter, this translates into a volume of 2.07 million cubic meters of anthracite per year. See Arbeitsgemeinschaft Energiebilanzen, *Energiebilanzen der Bundesrepublik Deutschland*, 2007, table 1.5.2.

35. Under certain circumstances, carbon dioxide can become a solid below 9°C. See Metz et al., *IPCC Special Report on Carbon Dioxide Capture and Storage*, p. 285.

36. See United Nations, *United Nations Convention on the Law of the Sea of 10 December 1982*, Division for Ocean Affairs and the Law of the Sea; European Commission, *Proposal for a Directive of the European Parliament and of the Council on the Geological Storage of Carbon Dioxide*, COM (2008) 18, January 23, 2008.

37. See Deutsche Energie-Agentur, *Energiewirtschaftliche Planung für die Netzintegration von Windenergie in Deutschland an Land und Offshore bis zum Jahr 2020*, Dena-Netzstudie, 2005, p. 12; Bundesministerium für Umwelt, Naturschutz und Reaktorsicherheit, *Entwicklung der erneuerbaren Energien im Jahr 2009*, 2010, p. 20.

38. L. Bölkow, *Energie im nächsten Jahrhundert* (Knoth, 1987).

39. Desertec Foundation, *Clean Power from Deserts, The DESERTEC Concept for Energy, Water and Climate Security*, fourth edition, 2009.

40. D. H. Meadows, D. L. Meadows, J. Randers, and W. W. Behrens, *The Limits to Growth* (Universe Books, 1972).

41. These and the following data stem from Bundesanstalt für Geowissenschaften und Rohstoffe, *Energierohstoffe 2009: Reserven, Ressourcen, Verfügbarkeit*, 2009.

42. This is not to say that this would suffice for their extraction to make economic sense, as that depends as well on what prices can be expected with later extraction that would be forfeited by extracting immediately.

43. Because current prices are significantly higher than this, the recoverable reserves and contingent resources increase accordingly. This was disregarded here for the sake of facilitating comparisons in operational life. See also International Atomic Energy Agency and OECD/NEA, *Uranium 2005: Resources, Production and Demand*, 2006.

44. "China meldet Durchbruch in nuklearer Wiederaufbereitung," *Frankfurter Allgemeine Zeitung*, January 4, 2011, p. 5; "The business of nuclear. China now reprocessing: A beginning, not a breakthrough," *Fuel Cycle Week*, January 6, 2011.

45. See T. Williams, *Kernreaktoren der nächsten Generation in Planung*, FLASH, October 2005, p. 8.

46. OECD Nuclear Energy Agency and International Atomic Energy Agency, *Uranium 2003: Resources, Production and Demand*, 2004, p. 22.

47. The cladding in the fusion reactor, exposed as it is to energetic radiation, does become somewhat radioactive. Tritium decays fairly quickly, with a half-life of 12.3 years. Waste disposal problems are very small relative to those for fission reactor waste.

48. For a review of the literature on marginal abatement costs, see R. S. J. Tol, "The marginal damage costs of carbon dioxide emissions: An assessment of the uncertainties," *Energy Policy* 33 (2005): 2064–2074.

49. *Building a Global Carbon Market*, Report pursuant to Article 30 of Directive 2003/87/EC COM(2006)676 final, Brussels, November 13, 2006.

50. R. H. Coase, "The problem of social cost," *Journal of Law and Economics* 3 (1960): 1–44.

51. European Commission, *EU Action Against Climate Change—EU Emissions Trading—An Open Scheme Promoting Global Innovation*, 2005.

52. International Energy Agency, CO_2 *Emissions from Fuel Combustion*, release 01, CO_2 Sectoral Approach, 2007 (http://www.sourceoecd.org).

53. European Commission, press release IP/06/1862, December 20, 2006.

54. Proposal for a directive of the European Parliament and of the Council amending Directive 2003/87/EC so as to improve and extend the greenhouse gas emission allowance trading system of the Community COM (2008) final, January 23, 2008.

55. Based on emissions from energy generation and industry in general. Source: http://dataservice.eea.europa.eu.

56. On April 19, 2006, the maximum stood at 29.95 euros per metric ton of CO_2.

57. E. Heymann, *EU-Emissionshandel—Verteilungskämpfe werden härter*, Deutsche Bank Research, Aktuelle Themen 377, Frankfurt a.M., January 25, 2007.

58. D. Dürr, *Der europäische Emissionshandel*, Eurostat website (http://epp .eurostat.ec.europa.eu), July 2008, including an EU estimate for 2008.

59. Proposal for a directive of the European Parliament and of the Council amending Directive 2003/87/European Commission, January 23, 2008.

60. Proposal for a directive of the European Parliament and of the Council amending Directive 2003/87/EC so as to improve and extend the greenhouse gas emission allowance trading system of the Community, COM (2008) final, January 23, 2008. Exceptions are allowed for community heating generation.

61. "Ein Teil des Geldes fliesst zurück," *Handelsblatt* (http://www.handelsblatt .com), August 19, 2008.

62. "Kalifornien schafft Emissionshandel," *Süddeutsche Zeitung* (http://www .sueddeutsche.de), December 18, 2010.

63. Council of the European Union, *Brussels European Council 8/9 March 2007—Presidency Conclusions*, 7224/1/07 REV 1 CONCL 1, Brussels, May 2, 2007.

64. M. Frondel, N. Ritter, C. M. Schmidt, and C. Vance, "Die ökonomischen Wirkungen der Förderung Erneuerbarer Energien: Erfahrungen aus Deutschland," *Zeitschrift für Wirtschaftspolitik* 59 (2010): 107–133.

65. Scientific Advisory Council at the Federal Ministry of Economics and Technology, *Zur Förderung erneuerbarer Energien*, Advisory Opinion, Dokumentation 534, Berlin 2004. See also M. Frondel, N. Ritter, and C. M. Schmidt: *Photovoltaik: Wo viel Licht ist, ist auch viel Schatten*, RWI, Positionen 18.2, Essen 2007; J. Weimann, *Die Klimapolitikkatastrophe. Deutschland im Dunkel der Energiesparlampe* (Metropolis, 2008), p. 49f.

66. C. Kemfert and J. Diekmann, "Förderung erneuerbarer Energien und Emissionshandel—wir brauchen beides," *DIW Wochenbericht* no. 11, 2009: 169–174.

Chapter 3

1. According to Einstein's energy formula, during photosynthesis minuscule particles of mass, far too tiny to be weighed even with the most sensitive instruments, are created out of the energy of sunlight, which can be transformed back into energy by combustion. The mass defect in this chemical reaction lies below the measurable threshold, like that occurring in nuclear reactions, which is why chemists rightly talk of the "law of conservation of mass". The law of conservation of mass is only plausible heuristics. Strictly speaking, according to the Special Theory of Relativity the mass of a molecule (or, for that matter, of an atom itself) is always smaller than the sum of the mass of its constituent atoms (or subatomic particles), because during the formation of the molecule binding energy will be released that arises from the mass that goes missing. See U. C. Harten, *Physik. Einführung für Ingenieure und Naturwissenschaftler* (Springer, 2003), p. 363f.

2. Biomass is defined as organic matter, including all living organisms, dead organisms, and organic products from metabolic processes. Fossil fuels aren't included. Microorganisms account for about 60% of the Earth's biomass.

3. United States District Court for the Northern District of California, Case C 05-0898 CRB, Memorandum and Order, filed August 22, 2006.

4. See, e.g., S. Pacala and R. Socolow, "Stabilization wedges: Solving the climate problem for the next 50 years with current technologies," *Science* 305 (2004): 968–972.

5. Scientific Advisory Council on Agricultural Policy at the German Federal Ministry of Food, Agriculture and Consumer Protection, *Nutzung von Biomasse zur Energiegewinnung, Empfehlungen an die Politik*, November 2007.

6. See the website of the Walter Thiel GmbH, July 2007 (http://www.thiel-heizoel.de).

7. International Energy Agency, *World Energy Outlook 2009*, 2009; International Energy Agency, *Renewables Information 2009*, 2009; German Federal Ministry for the Environment, Nature Conservation and Nuclear Safety,

Time Series for the Development of Renewable Energies in Germany (http://erneuerbar.info).

8. See Renewable Fuels Association, *2010 Ethanol Industry Outlook: Climate of Opportunity* (www.ethanolrfa.org); Emerging Markets Online, *Biodiesel 2020,* second edition (www.emerging-markets.com).

9. See US Department of Agriculture, *World Agricultural Supply and Demand Estimates,* January 12, 2011. According to the Department of Agriculture's estimate, 4.55 billion bushels of corn for ethanol were produced in 2009–10, which is 35% of the total production of 13 billion bushels of corn in 2009–10. If equal productivities are assumed, this means that 35% of a total area of 322,000 square kilometers used for corn crops—113,000 square kilometers—will be used for ethanol production. The forecast for 2010–11 is that 4.9 billion bushels of corn will be used for ethanol production in the year, which is 39.4% of the total production of 12.45 billion bushels of corn in 2010/11. This corresponds to 356,900 square kilometers devoted to corn crops, of which 140,600 square kilometers are used for ethanol production.

10. See EurObserv'ER, *Biofuels Barometer,* July 2010 (http://www.eurobserv-er .org).

11. See European Biodiesel Board, *Statistics,* 2008 (http://www.ebb-eu.org); European Union of Ethanol Producers, *2006 EU Fuel Ethanol—Production vs Consumption,* 2007 (www.uepa.be).

12. See International Energy Agency, *Energy Balances of OECD Countries,* 2010, p. II.240.

13. See "Biofuels: Prospects, risks and opportunities," in *The State of Food and Agriculture* (Food and Agriculture Organization of United Nations, 2008), p. 30f.; US Department of Energy Office of Science, *Biofuels Policy and Legislation,* last modified April 19, 2010 (http://genomicscience.energy.gov).

14. See "Biofuels: Prospects, risks and opportunities," pp. 24–25.

15. Biokraftstoffquotengesetz—BioKraftQuG, December 18, 2006, Bundestagsdrucksache 16/2709.

16. EU Directive 2009/28/EC of the European Parliament and of the Council of 23 April 2009 on the Promotion of the Use of Energy from Renewable Sources and Amending and Subsequently Repealing Directives 2001/77/EC and 2003/30/EC.

17. According to the Sprengel rule, plants need all their nutrients in certain proportions. If less of one nutrient is available, the plant grows more slowly, even if the other nutrients are available in excess. And when a scarcer nutrient becomes available in larger quantities, the plant grows only in proportion to the amount of the other nutrients available. If the natural nutrient distribution in the soil is non-uniform, it is then necessary to add the missing nutrients in each case in order to ensure optimum growth of all plants and avoid nutrient excesses.

18. T. W. Patzek, "The real biofuel cycles," manuscript, University of California, Berkeley, July 11, 2006.

19. P. J. Crutzen, A. R. Mosier, K. A. Smith, and W. Winiwarter, "N_2O release from agro-biofuel production negates global warming reduction by replacing fossil fuels," *Atmospheric Chemistry and Physics* 7 (2007): 11191–11205, and (with the same title) 8 (2008): 389–395. Similarly critical conclusions are drawn in the literature presented later in this chapter.

20. A. J. Liska, H. S. Yang, V. R. Bremer, T. J. Klopfenstein, D. T. Walters, G. E. Erickson, and K. G. Cassman, "Improvements in life cycle energy efficiency and greenhouse gas emissions of corn-ethanol," *Journal of Industrial Ecology* 13 (2008): 58–74.

21. In eastern Germany, Choren Industries GmbH, a company partly owned by Volkswagen and Daimler, and temporarily by Shell, operates a pilot plant built to produce synthetic diesel fuel from biomass using the Carbo-V process. In January 2010, Choren announced a joint venture with the French company CNIM to build a BtL diesel-fuel plant based on the gasification technologies developed by Choren, with a capacity of 23,000 tons per year. Production is to start in 2014. Subsidiaries of Choren operate in the United States, in China, and in Malaysia.

22. See Sächsische Landesanstalt für Landwirtschaft, *Biodieseleinsatz—FAME oder RME?*, Newsletter, Dresden, November 30, 2007 (http://www.smul.sachsen. de).

23. For a more critical view on this process, see Wissenschaftlicher Beirat Agrar-politik, *Nutzung von Biomasse zur Energiegewinnung* (Federal Government of Germany, 2007), p. 182. Although the Council endorsed the method, it objected that not all biogenic waste should be used to produce fuel, as returning some of this waste to the fields fulfills an important ecological function (humus creation, nutrient return). If straw is used to produce diesel fuel, the nutrients it usually delivers (phosphorus, calcium, magnesium, potassium) will end up as slag, which can be processed into fertilizer only with much effort.

24. See L. A. Martinelli and S. Filoso, "Expansion of sugarcane ethanol production in Brazil: Environmental and social challenges," *Journal of Applied Ecology* 18 (2008): 885–898.

25. This and the following numbers were calculated on the basis of data from J. Fargione, J. Hill, D. Tilman, S. Polasky, and P. Hawthorne, "Land clearing and the biofuel carbon debt," *Science* 319 (2008): 1235–1238, here from the data shown in columns 3 and 4 of figure 1. For similar calculations, see T. Searchinger, R. Heimlich, R. A. Houghton, F. Dong, A. Elobeid, J. Fabiosa, S. Tokgoz, D. Hayes, and T. H. Yu., "Use of US croplands for biofuels increases greenhouse gases through emissions from land-use change," *Science* 319 (2008): 1238–1240; H. K. Gibbs, J. Jonston, J. A. Foley, T. Holloway, C. Monfreda, N. Ramankutty, and D. Zaks, "Carbon payback times for tropical biofuel expansion: The effects of changing yield and technology," *Environmental Research Letters* 3 (2008), 034001, 200.

26. For a discussion and an overview of the literature, see A. Young, "Is there really spare land? A critique of estimates of available cultivable land in

developing countries," *Development and Sustainability* 1 (1999): 3–18; T. Beringer, W. Lucht, and S. Schaphoff, "Bioenergy production potential of global biomass plantations under environmental and agricultural constraints," manuscript, Potsdam Institute for Climate Impact Research, forthcoming in *Global Change Biology Bioenergy.*

27. The IEA assumes that 14 million hectares (35 million acres) would be needed for a worldwide 1.0% biofuel share. This corresponds to a yield of 1.5 metric tons of oil-equivalent per hectare. Biofuels accounted for 0.8% of fuels in the EU 15 countries in 2005, equivalent to 2.6 million metric tons oil-equivalent, which, according to the average yields reported by the IEA, require cropland to the tune of 1.7 million hectares. Scaled up to a biofuel share of 10%, the necessary area would be 21.5 million hectares, equivalent to 31% of the entire arable land in the EU 15. The figures for a 20% and a 100% biofuel share were arrived at by simply scaling up the above numbers.

28. In the US, 2 million hectares (5 million acres) were needed in 2000 for a 1.6% biofuel share, equivalent to 3.1 million metric tons oil equivalent, which corresponds to a yield of 1.6 metric tons of oil equivalent per hectare. This required 2 million hectares of cropland. Scaled up to a biofuel share of 10%, the necessary area would be 12.5 million hectares, equivalent to 9% of the arable land in the US. The figures for a 20% and a 100% biofuel share again result from scaling up the above numbers.

29. European Commission, *The Impact of Minimum 10% Obligation for Biofuel Use in the EU-27 in 2020 on Agricultural Markets*, AGRI G-2/WM D. DG Agri, Brussels 2007.

30. B. Dehue, S. Meyer, and W. Hettinga, *Review of EU's Impact Assessment of 10% Biofuels on Land Use Change,* prepared for Gallagher Commission Review 2008, Ecofys b.v., The Netherlands. See also B. Eickhout, G. J. van den Born, J. Notenboom, M. van Oorschot, J. P. M. Ros, D. P. van Vuuren, and H. J. Westhoek, *Local and Global Consequences of the EU Renewable Directive for Biofuels,* MNP Report 50143001, 2008, Environmental Assessment Agency, The Netherlands.

31. A. Hoffmann and P. Liebrich, "Magere Ernte verteuert Brot und Bier," *Süddeutsche Zeitung*, August 20, 2007 (http://www.sueddeutsche.de).

32. Fachagentur Nachwachsende Rohstoffe, *Daten und Fakten* (http://www.fnr-server.de); European Commission, DG Agriculture and Rural Development, *Bioenergy* (http://ec.europa.eu); European Biomass Industry Association (www.eubia.org); Bundesministerium für Ernährung, Landwirtschaft und Verbraucherschutz, *Statistisches Jahrbuch über Ernährung, Landwirtschaft und Forstender Bunderepublik Deutschland 2009* (Wirtschaftsverlag NW, 2009).

33. US Department of Agriculture, *World Corn Production, Consumption and Stocks* (http://www.fas.usda.gov) and *Grain: World Markets and Trade—World Corn Situation: US. Expected to Continue to Dominate Market* (http://www.fas.usda.gov).

34. World corn production rose during this period by 55 million tons, while US consumption of corn for bioethanol production rose by 50 million tons (D. Mitchell, *A Note on Rising Food Prices*, World Bank Working Paper 4682, 2008).

35. J. von Braun, *High Food Prices: The What, Who and How of Proposed Policy Actions*, Policy Brief, International Food Policy Research Institute, 2008; J. von Braun, "Unbezahlbare Nahrungsmittel—stark gestiegene Nachfrage oder Agrarrohstoffe als Anlageklasse—was sind die Ursachen?" *ifo Schnelldienst* 61 (2008), no. 11: 3–6.

36. Energy Tax Law of 1978, Public Law 95-618, 95th Congress, November 9, 1978.

37. Energy Policy Act of 2005, Public Law 109-58, 109th Congress, August 8, 2005. See also Clean Fuels Development Coalition, *The Ethanol Fact Book—A Compilation of Information About Fuel Ethanol* (http://www.ethanol.org).

38. Energy Independence and Security Act of 2007, Public Law 110-140, 110th Congress, December 19, 2007.

39. Food, Conservation, and Energy Act of 2008, Public Law 110–234, 110th Congress, May 22, 2008.

40. M. W. Rosegrant of International Food Policy Research Institute, "Biofuels and grain prices: Impacts and policy responses," testimony before US Senate Committee on Homeland Security and Governmental Affairs, May 7, 2008.

41. See Mitchell, "A note on rising food prices." A preliminary version of the World Bank working paper (April 8, 2008) that had circulated among a few scholars gave rise to a heated controversy that caught the eye of the press. It attributed three-fourths of the price increase for food to the withdrawal of arable land for energy crops. The official version, released in July as World Bank Working Paper 4682, states only that "most" of the 70–75% of the price increase that isn't accounted for by other causes resulted from land withdrawal for energy crops, a slight but interesting change in wording.

42. See, e.g., K. Collins, The Role of Biofuels and other Factors in Increasing Farm and Food Prices: A Review of Recent Developments with a Focus on Feed Grain Markets and Market Prospects, report commissioned by Kraft Food Global, June 19, 2008 (http://www.globalbioenergy.org), or J. Lipsky, "Commodity prices and global inflation," remarks by the First Deputy Managing Director of the IMF to Council on Foreign Relations, New York, May 8, 2008.

43. J. Piesse and C. Thirtle, "Three bubbles and a panic: An explanatory review of recent food commodity price events," *Food Policy* 34 (2009): 119–129.

44. D. Headey and S. Fan, "Anatomy of a crisis: The causes and consequences of surging food prices," *Agricultural Economics* 39 (2008), supplement: 375–391.

45. C. L. Gilbert, "How to understand high food prices," *Journal of Agricultural Economics*. 61 (2010): 398–425. See also J. Baffes and T. Haniotis, *Placing the 2006/08 Commodity Price Boom into Perspective*, World Bank Working Paper

5371, 2010; A. Ajanovic, "Biofuels versus food production: Does biofuels production increase food prices?" *Energy* 36 (2010): 2070–2076.

46. See H.-W. Sinn, *Casino Capitalism* (Oxford University Press, 2010), chapter 1.

47. H.-W. Sinn, "Tanken statt essen?" *WirtschaftsWoche*, September 3, 2007, p. 162, and *ifo Standpunkt* no. 88, 2007 (http://www.ifo.de).

48. World Bank, PovcalNet Online Poverty Analysis Tool (http://www .worldbank.org).

49. M. Böhm, *Bayerns Agrarproduktion 1800–1870* (Scripta Marcarturae, 1995); see especially pp. 403–407.

50. G. Höher, "Energiepflanzenanbau in Niedersachsen: Aktueller Stand und Perspektiven," lecture given at Third Biogas-Fachkonkress, Hitzacker, 2008.

51. T. R. Malthus, *An Essay in the Principle of Population*, 1798 (Penguin, 1985).

52. A. Smith, *An Inquiry into the Nature and Causes of the Wealth of Nations*, 1776 (Prometheus Books, 1991), p. 84.

Chapter 4

1. Brown coal contains 70% carbon and 5.5% hydrogen. Anthracite contains 93% carbon and 3% hydrogen. The rest consists of oxygen, nitrogen, and sulfur. The other substances that occur with carbon, such as water, ash, silicon, potassium, and aluminum, have been disregarded.

2. The 30.7% value mentioned refers to the energy directly available, the (lower) calorific value. The upper calorific value of a hydrogen atom amounts to 36.3% of the energy of a carbon atom.

3. The analysis provided in this chapter and the next is based largely on my Presidential Address at the Annual Congress of the International Institute of Public Finance in Warwick in August 2007 and my Thünen Lecture at the Annual Congress of the Verein für Socialpolitik in Munich in October 2007. For the published versions, see "Public policies against global warming: A supply side approach," *International Tax and Public Finance* 15 (2008): 360–394; "Das grüne Paradoxon: Warum man das Angebot bei der Klimapolitik nicht vergessen darf," *Perspektiven der Wirtschaftspolitik* 9 (2008): 109–142. Earlier studies that consider the supply side can be divided into two kinds: static general equilibrium models (mostly of the GTAP type) that consider a resource market common to all countries, and analytical intertemporal models with a single sector that examine the gradual depletion of directly consumable (i.e., not used for production) fossil-fuel resources in conjunction with the simultaneous accumulation of pollutants in the atmosphere. On studies of the first kind, see R. Gerlagh and O. Kuik, *Carbon Leakage with International Technology Spillovers*, Fondazione Eni Enrico Mattei Working Paper 33, 2007. On the second kind, see J. A. Krautkraemer, "On growth, resource amenities and the preservation of natural

environments," *Review of Economic Studies* 52 (1985): 153–170; C. D. Kolstad and J. A. Krautkraemer, "Natural resource use and the environment," in *Handbook of Natural Resource and Energy Economics*, volume 3, ed. A. V. Kneese and J. L. Sweeney (Elsevier, 1993); C. Withagen, "Pollution and exhaustibility of fossil fuels," *Resource and Energy Economics* 16 (1994): 235–242; M. Hoel and S. Kverndokk, "Depletion of fossil fuels and the impacts of global warming," *Resource and Energy Economics* 18 (1996): 115–136; Y. H. Farzin, "Optimal pricing of environmental and natural resource use with stock externalities," *Journal of Public Economics* 62 (1996): 31–57; O. Tahvonen, "Fossil fuels, stock externalities, and backstop technology," *Canadian Journal of Economics* 30 (1997): 855–874; A. Krautkraemer, "Non-renewable resource scarcity," *Journal of Economic Literature* 36 (1998): 2065–2107. The approach chosen here links the country approach with an intertemporal view. It goes beyond the intertemporal models mentioned in that it considers the consumption of the resource not as an activity that provides utility directly but as a production factor for the generation of the domestic product, and in that it analyzes the social portfolio problem in the choice between man-made and natural capital with a view to the climate problem.

4. N. Stern, *The Economics of Climate Change—The Stern Review* (Cambridge University Press, 2007).

5. What I call "counterfactual history" here is called "comparative-static analysis" in the jargon of economics.

6. See S. Felder and T. F. Rutherford, "Unilateral CO_2 reductions and carbon leakage: The consequences of international trade in oil and basic materials," *Journal of Environmental Economics and Management* 25 (1993): 162–176; J.-M. Burniaux and J. Oliveira Martins, *Carbon Emission Leakages: A General Equilibrium View*, OECD Working Paper 242, 2000.

7. Governments, however, will face a moral question when it comes to deciding whether to accept commitments under the Kyoto succession agreement. This will be discussed in chapter 5.

8. This is the so-called Carbon-Leakage Hypothesis. See M. Hoel, "Should a carbon tax be differentiated across sectors?" *Journal of Public Economics* 59 (1996): 17–32, or O. J. Kuik and R. Gerlagh, "Trade liberalization and carbon leakage," *The Energy Journal* 24 (2003): 97–120.

9. Deutsche Presse-Agentur, January 16, 2008.

10. A. Endres, "Tanz um die Tonne—Fünf Fragen und Antworten zum Ölpreis," *Die Zeit*, April 24, 2008 (http://www.zeit.de); Bundesamt für Geowissenschaften und Rohstoffe, *Reserven, Ressourcen und Verfügbarkeit von Energierohstoffen 2006* (http://www.bgr-bund.de).

11. D. H. Meadows, D. L. Meadows, J. Randers, and W. W. Behrens, *The Limits to Growth* (Universe Books, 1972).

12. The concepts of reserves and resources aren't uniformly defined in the literature. Sometimes resources are defined as excluding reserves. In resource economics, the prevailing definition is the expanded one as used in this book.

13. Methane hydrate is stable only at high pressures and low temperatures. It can be mined by reducing pressure, a process that is particularly suitable in permafrost areas. Mining of ocean stocks usually requires pumping heat. It is also possible to melt hydrates by injecting methanol.

14. These numbers refer to the carbon content, not the energy content.

15. See F. S. Chapin III, P. A. Matson, and H. A. Mooney, *Principles of Terrestrial Ecosystem Ecology* (Springer, 2002), p. 335f.

16. D. Archer, "Fate of fossil fuel CO_2 in geologic time," *Journal of Geophysical Research* 110 (2005): 5–11; D. Archer and V. Brovkin, "Millennial atmospheric lifetime of anthropogenic CO_2," *Climate Change*, unpublished manuscript, 2006; G. Hoos, R. Voss, K. Hasselmann, E. Meier- Reimer, and F. Joos, "A nonlinear impulse response model of the coupled carbon cycle-climate system (NICCS)," *Climate Dynamics* 18 (2001): 189–202.

17. Archer, "Fate of fossil fuel CO_2 in geologic time"; Archer and Brovkin, "Millennial atmospheric lifetime of anthropogenic CO_2."

18. H. Bachmann, *Die Lüge der Klimakatastrophe. Das gigantischste Betrugswerk der Neuzeit. Manipulierte Angst als Mittel zur Macht*, fourth edition (Frieling, 2008), p. 85.

19. See K. Trenberth, "Seasonal variations in global sea-level pressure and the total mass of the atmosphere," *Journal of Geophysical Research—Oceans and Atmospheres* 86 (1981): 5238–5246.

20. Since 1850, about 136 gigatons of carbon have been released through changes in land use. See R. T. Watson, I. R. Noble, B. Bolin, N. H. Ravindranath, D. J. Verardo, and D. J. Dokken, eds., *Land Use, Land-Use Change, and Forestry* (Cambridge University Press, 2000), p. 4. In this way, about 61 gigatons of carbon were released into the atmosphere by changes in land use, which, together with the 156 gigatons (347 GtC × 0.45), corresponds to approximately 200 gigatons of carbon.

21. H. Hotelling, "The economics of exhaustible resources," *Journal of Political Economy* 39 (1931): 137–175.

22. See. e.g., Sinn, "Public policies against global warming."

23. See J. E. Stiglitz, "Monopoly and the rate of extraction of exhaustible resources," *American Economic Review* 66 (1976): 655–661. (In the absence of extraction costs a monopolist chooses an extraction path such that the marginal revenue from sales of the resource increases at a rate that equals the market rate of interest. In the simplest case of constant demand elasticity, marginal revenue is proportional to the price, however, so that the Hotelling Rule applies just as in the competition case. Non-constant demand elasticity modifies the Hotelling Rule, but the modification can go in either direction.) Note, however, that the Hotelling Rule always applies to the stock in situ.

24. K. J. Arrow and G. Debreu, "Existence of an equilibrium for a competitive economy," *Econometrica* 22 (1954): 265–290.

25. In the case of renewable resources, sustainability is defined in such a way that the stock of this resource remains intact. This is not possible for exhaustible resources. Therefore, sustainability can be understood in this case only as a measured exploitation of the resource.

26. This argument was introduced to economic theory in a different context by Irving Fisher. It is known as the Fisher Separation Theorem. See I. Fisher, *The Rate of Interest: Its Nature, Determination and Relation to Economic Phenomena* (Macmillan, 1907).

27. See T. Page, *Conservation and Economic Efficiency. An Approach to Materials Policy* (Johns Hopkins University Press, 1977); R. M. Solow, "The economics of resources or the resources of economics," *American Economic Review* 64 (1974): 1–14; S. Anand and A. Sen, "Human Development and economic sustainability," *World Development* 28 (2000): 2029–2049.

28. See Stern, *The Economics of Climate Change*, especially the annex to chapter 2. William Nordhaus has observed that the high damages from the greenhouse effect calculated by the Stern Review can be attributed significantly to the assumption of a very low discount rate. See W. D. Nordhaus, "A review of the Stern Review on the economics of climate change," *Journal of Economic Literature* 45 (2007): 686–702.

29. R. M. Solow, "Intergenerational equity and exhaustible resources," *Review of Economic Studies* 41 (1974): 29–45; J. E. Stiglitz, "Growth with exhaustible natural resources: Efficient and optimal growth paths," *Review of Economic Studies* 41 (1974): 123–137. For the generalization of the efficiency condition in the case of stock-dependent extraction costs, see H.-W. Sinn, "Stock-dependent extraction costs and the technological efficiency of resource depletion," *Zeitschrift für Wirtschafts- und Sozialwissenschaften* 101 (1980): 507–517, where the generalization attempt by Geoffrey Heal is refuted. See G. Heal, "Intertemporal allocation and intergenerational equity," in *Erschöpfbare Ressourcen, Berichte von der Jahrestagung des Vereins für Socialpolitik 1979*, ed. H. Siebert (Duncker und Humblot, 1980).

30. Literally: "Climate change is the greatest market failure the world has ever seen. . . ." See Stern, *The Economics of Climate Change*, p. VIII.

31. H.-W. Sinn, *Pareto Optimality in the Extraction of Fossil Fuels and the Greenhouse Effect*, CESifo Working Paper 2083, 2007; NBER Working Paper 13453, 2007. These linkages are overlooked by Björn Lomborg, who asserts that measures to curb CO_2 emissions aren't worth it, because they just postpone the damage (B. Lomborg, *The Skeptical Environmentalist: Measuring the Real State of the World*, Cambridge University Press, 1998, pp. 258–324). See also W. D. Nordhaus, *Managing the Global Commons: The Economics of Climate Change* (MIT Press, 1994). Nordhaus presents an intertemporal computer model with climate change. His book, however, doesn't contain an analysis comparable in any way to this one, and it discusses the economic mechanisms operating in his model only in rudimentary fashion.

32. For the precise derivation of this condition, see Sinn, *Pareto Optimality*. See also Sinn, "Public Policies."

33. N. V. Long, "Resource extraction under the uncertainty about possible nationalization," *Journal of Economic Theory* 10 (1975): 42–53; K. A. Konrad, T. E. Olson, and R. Schöb, "Resource extraction and the threat of possible expropriation: The role of Swiss bank accounts," *Journal of Environmental Economics and Management* 26 (1994): 149–162.

Chapter 5

1. The phenomenon was first described in newspaper articles. See, e.g., H.-W. Sinn, "The green paradox," *Project Syndicate*, June 2007, published in *Journal of Turkish Weekly* (Turkey), *Les Nouvelles* (Madagascar), *Les Echos* (Mali), *Standard Times* (Sierra Leone), *South China Morning Post* (Hong Kong), *The Financial Express* (India), *The Korea Herald* (South Korea), *Business World* (Philippines), *The Sunday Times* (Sri Lanka), *The Nation* (Thailand), *Die Presse* (Austria), *L'Echo* (Belgium), *Borsen* (Danmark), *Aripaev* (Estonia), *Vilaggazdasag* (Hungary), *The Times of Malta* (Malta), *Danas* (Serbia), *Stabroek News* (Guyana), *Jordan Property* (Jordan), *Al Raya* (Qatar), *Al Eqtisadiah* (Saudi Arabia), *Duowei Times* (United States); H.-W. Sinn, "Die Logik der Scheichs," *Die Welt*, July 9, 2007; H.-W. Sinn, "Greenhouse gases: Demand control policies, supply and the time path of carbon prices," *Vox*, October 31, 2007 (http://www.vox.org). More formal, scholarly discussions followed: H.-W. Sinn, "Public policies against global warming: A supply side approach," *International Tax and Public Finance* 15 (2008): 360–394 (presidential address, World Congress, International Institute of Public Finance, Warwick, August 2007, CESifo Working Paper 2087, 2007); H.-W. Sinn, "Das grüne Paradoxon: Warum man das Angebot bei der Klimapolitik nicht vergessen darf," *Perspektiven der Wirtschaftspolitik* 9 (2008): 109–142 (Thünen Lecture, Annual Meeting, Verein für Socialpolitik, Munich, September 2007). The first German edition of the present book was *Das grüne Paradoxon* (Econ, 2008); the second followed a year later. In addition I wrote quite a number of smaller pieces for various newspapers and magazines. They are listed, along with the reactions of German politicians, in my résumé at www.cesifo.org/hws. The first scholarly discussions of my views were in German; they included the following: A. Endres, "Ein Unmöglichkeitstheorem für die Klimapolitik?" *Perspektiven der Wirtschaftspolitik* 9 (2008): 350–382; R. S. J. Tol and D. Anthoff, "Kommentar zu Hans-Werner Sinn: Kämpft alle Klimapolitik mit dem Grünen Paradoxon?" in *Diskurs Klimapolitik, Jahrbuch Ökologische Ökonomik 6*, ed. F. Beckenbach et al. (Metropolis, 2009); O. Edenhofer and M. Kalkuhl, "Kommentar zu Hans-Werner Sinn: Das 'Grüne Paradoxon'—Menetekel oder Prognose," in ibid. Meanwhile, a swelling flow of scholarly papers discussing the Green Paradox have come out in English, some of which will be cited in the following pages. Meanwhile, the term "green paradox" has also been applied to a somewhat different phenomenon, namely the precautionary saving,

capital accumulation, and environmentally damaging growth induced by people's attempt to compensate for announced future energy price increases and the poverty resulting from it. See S. Smulder, Y. Tsur, and A. Zemel, "Announcing climate policy: Can a green paradox arise without scarcity?" (presented at World Congress of Environmental and Resource Economists, Montreal, 2010).

2. While Vattenfall's east German sites cover only one-tenth of Germany's brown coal reserves, the company accounts for one-third of Germany's current brown coal extraction.

3. For the formal dynamic optimization model on which these considerations are based, see Sinn, "Public policies against global warming."

4. For a full derivation within the framework of a formal intertemporal market model, see "Public policies against global warming," in which I show that the neutrality condition for the constancy of the present value of the (absolute) price wedge ΔP is equivalent to the condition that the relative price wedge $\Delta P/P$ increase over time at a rate equal to the product of the discount rate (interest plus, possibly, expropriation probability) and the share of extraction costs in the proceeds from the sale of the resources. For a non-technical discussion of this topic, see "Das grüne Paradoxon," in *Perspektiven der Wirtschaftspolitik*; also see "Global warming: The neglected supply side," in *The EEAG Report on the European Economy* (CESifo, 2008). The theoretical core of these articles lies in a formal theorem on the effects of price changes on the resource-extraction pathway derived in N. V. Long and H.-W. Sinn, "Surprise price shifts, tax changes and the supply behaviour of resource extracting firms," *Australian Economic Papers* 24 (1985): 278–289. This theorem, in turn, is the generalization of an intertemporal equilibrium model used to study the effects of non-constant unit and *ad valorem* tax rates on the extraction of fossil fuels; see H.-W. Sinn, "Absatzsteuern, Ölförderung and das Allmendeproblem," in *Reaktionen auf Energiepreisänderungen*, ed. H. Siebert (Lang, 1982).

5. T. Eichner and R. Pethig have extended the corresponding analysis of my German book to an intertemporal general equilibrium model with endogenous goods prices. See *Carbon Leakage, the Green Paradox and Perfect Future Markets*, CESifo Working Paper 2542, 2009 (forthcoming in *International Economic Review*).

6. Proposal for a directive of the European Parliament and of the Council amending Directive 2003/87/EC so as to improve and extend the greenhouse gas emission allowance trading system of the Community COM(2008) 16 final—2008/0013 (COD), January 23, 2008.

7. See the chapter titled "On the optimal order of exploitation of deposits of an exhaustible resource" in M. Kemp and N. V. Long, *Exhaustible Resources, Optimality and Trade* (North-Holland, 1980),

8. If there is an important cost component in this price, it is the so-called user cost. The user cost is the present value of a price at which a resource could be sold in the future and which therefore has to be relinquished if the resource is sold today.

9. For alternative information on extraction costs with even lower figures, see *Resources to Reserves* (International Energy Agency, 2005, p. 110f); E. Harks, *Der globale Ölmarkt—Herausforderungen and Handlungsoptionen für Deutschland* (Stiftung Wissenschaft und Politik, Deutsches Institut für Internationale Politik und Sicherheit, 2007), p. 11; C. Jojarth, *The End of Easy Oil: Estimating Production Costs for Oil Fields around the World*, Working Paper 72, Stanford University Program on Energy and Sustainable Development, 2008.

10. In Sinn, "Public policies," this result follows from the assumption that the price elasticity of demand and the unit extraction cost remain bounded, i.e., can't become infinite as the remaining resource stock *in situ* and demand tend to zero.

11. In Germany the cost of solar power is currently about 40 euro cents per kilowatt-hour. The wholesale price of electricity is about 5 euro cents per kilowatt-hour.

12. For detailed conditions under which the Green Paradox may or may not arise, even when replacement technologies impose a price ceiling below the unit extraction costs of some remnants of high-cost reserves, see R. Q. Grafton, T. Kompas, and N. V. Long, *Biofuels and the Green Paradox*, CESifo Working Paper 2960, 2010; F. van der Ploeg and C. Withagen, *Is There Really a Green Paradox?* CESifo Working Paper 2963, 2010; R. Gerlagh, "Too much oil," *CESifo Economic Studies* 56 (2010): 79–102.; M. Hoel, *Bush Meets Hotelling: Effects of Improved Renewable Energy Technology on Greenhouse Gas Emissions*, CESifo Working Paper 2492, 2008; Sinn, "Public policies"; Sinn, "Das grüne Paradoxon."

13. This argument was made by the German minister of the environment, S. Gabriel, in "Kurzarbeit im Elfenbeinturm" (*Handelsblatt*, June 4, 2009, p. 8), which was a reply to H.-W. Sinn, "Kurzarbeit auf den Bohrinseln" (*Handelsblatt*, May 28, 2009, p. 7). (For an article in English similar to the latter, see H.-W. Sinn, "How to resolve the green paradox," *Financial Times*, August 27, 2009.) For my answer to Gabriel's article, see "Kurzarbeit im Umweltministerium," *Handelsblatt*, July 20, 2009, p. 6.

14. Short-term and long-term demand and price changes aren't differentiated in the literature on carbon leakage in countries linked through the resource markets, because that literature originates from the area of the computable general equilibrium models (mostly of the Global Trade Analysis Project type), which are inherently static and don't consider intertemporal dynamics. See, e.g., B. J.-M. Burniaux and J. Oliveira Martins, *Carbon Emission Leakages: A General Equilibrium View*, OECD Working Paper 242, 2000, or R. Gerlagh and O. Kuik, *Carbon Leakage with International Technology Spillovers*, Working Paper 33, Fondazione Eni Enrico Mattei, 2007.

15. See E. Fehr and K. Schmidt, "The economics of fairness, reciprocity and altruism—Experimental evidence and new theories," in *Handbook of the Economics of Giving, Reciprocity and Altruism*, volume 1, ed. S.-C. Kolm and J. M. Ythier (Elsevier, 2006).

16. German Chancellor Angela Merkel accepted this position during a visit to China. See REGIERUNGonline, "Gutes Klima in Peking," August 27, 2007 (http://www.bundesregierung.de). Also see Chancellor Merkel's speech at the symposium "Deutschland und Japan—in gemeinsamer Verantwortung für die Zukunft," Kyoto, 2007 (http://www.bundeskanzlerin.de).

17. See K.-M. Mäler, "International environmental problems," *Oxford Review of Economic Policy* 6 (1990): 80–108; M. Hoel, "Global environmental problems: The effects of unilateral actions taken by one country," *Journal of Environmental Economics and Management* 20 (1991): 55–70.

18. W. Buchholz and K. A. Konrad, "Global environmental problems and the strategic choice of technology," *Journal of Economics* 60 (1994): 299–321. Compare *Klimapolitik zwischen Emissionsvermeidung und Anpassung* (Bundesministerium der Finanzen, 2010) (http://www.bundesfinanzministerium.de).

19. A. Endres and M. Finus, "Playing a better global warming game: Does it help to be green?" *Swiss Journal of Economics and Statistics* 134 (1998): 21–40. See also S. Barrett, "International environmental agreements as games," in *Conflicts and Cooperation in Managing Environmental Resources*, ed. R. Pethig (Springer, 1992); A. Endres, "Ein Unmöglichkeitstheorem für die Klimapolitik?" *Perspektiven der Wirtschaftspolitik* 9 (2008): 350–382; A. Endres, "Radfahren statt Trittbrettfahren?—Eine spieltheoretisch geleitete Einschätzung," *ifo Schnelldienst* 60 (2007), no. 7: 9–11; R. Pethig, "Bedingungen für den Erfolg internationaler Umweltabkommen ungünstig," *ifo Schnelldienst* 60 (2007), no. 7: 15–18.

20. This verbal formulation is heuristic to make it easier for the reader to follow. The formal proof of the marginal conditions for the social optimum assumes that the fossil fuels are a production factor that, together with capital and labor, explains the economy's output. See Sinn, "Public policies"; Sinn, *Pareto Optimality in the Extraction of Fossil Fuels and the Greenhouse Effect*, CESifo Working Paper 2083 and NBER Working Paper 13453, 2007.

21. *Model Double Taxation Convention on Income and on Capital* (OECD Committee on Fiscal Affairs, 1977).

22. See H.-W. Sinn, *The New Systems Competition* (Blackwell, 2003), chapter 2.

23. A. C. Pigou, *The Economics of Welfare* (Macmillan, 1920).

24. N. Stern, *The Economics of Climate Change—The Stern Review* (Cambridge University Press, 2007), p. 277, 362–364, 386, 532f.

25. See M. Hoel and S. Kverndokk, "Depletion of fossil fuels and the impacts of global warming," *Resource and Energy Economics* 18 (1996): 115–136; M. Kalkuhl and O. Edenhofer, "Prices versus quantities and the intertemporal dynamics of the climate rent," unpublished paper, Potsdam Institute for Climate Impact Research, 2010. The latter also try various versions of a proxy Pigouvian tax on the extraction flow, with the tax rate depending on individual or aggregate carbon accounts, but in the end come up with a negative conclusion about

meaningful carbon taxation. See also Edenhofer and Kalkuhl, "Kommentar zu Hans-Werner Sinn: Das 'Grüne Paradoxon'."

26. Only in a few special and unrealistic cases can the tax rate be described by a simple formula. For example, when the marginal damage and the interest rate are constant, the present value of the marginal damage is the quotient of the marginal damage and the interest rate, following the formula for the present value of a permanent annuity. See J. Strand, "Optimal taxation of an exhaustible resource with stock externalities, backstop technology and rising extraction costs," Tax Policy Division, Fiscal Affairs Department, International Monetary Fund, 2007.

27. H.-W. Sinn, "Optimal resource taxation," in *Risk and the Political Economy of Resource Development*, ed. D. Pearce, H. Siebert, and I. Walter (Macmillan, 1984).

28. For the implications of tax rate changes on the speed of resource extraction and/or capital accumulation, see Sinn, "Absatzsteuern"; Long and Sinn, "Surprise price shifts"; P. Howitt and H.-W. Sinn, "Gradual reforms of capital income taxation," *American Economic Review* 79 (1989): 106–124; P. J. N. Sinclair, "On the optimum trend of fossil fuel taxation," *Oxford Economic Papers* 46 (1994): 869–877; A. Ulph and D. Ulph, "The optimal time path of a carbon tax," *Oxford Economic Papers* 46 (1994): 857–868; E. R. Amundsen and R. Schöb, "Environmental taxes on exhaustible resources," *European Journal of Political Economy* 15 (1999): 311–329.

29. See Sinn, "Absatzsteuern," equation 14 and figure 1.

30. H.-W. Sinn, "Public policies," equation A17.

31. Commission of the European Communities, *European Economy: The Climate Challenge: Economic Aspects of the Community's Strategy for Limiting CO_2 Emissions*, 1992.

32. S. Bach, M. Kolhaas, V. Meinhardt, B. Praetorius, H. Wessels, and R. Zwiener, *Wirtschaftliche Auswirkungen einer ökologischen Steuerreform*, Sonderheft 153, Deutsches Institut für Wirtschaftsforschung, 1995, p. 212.

33. Kalkuhl and Edenhofer make this point in "Prices versus quantities and the intertemporal dynamics of the climate rent."

34. F. S. Chapin III, P. A. Matson, H. A. Mooney, *Principles of Terrestrial Ecosystem Ecology* (Springer, 2002), p. 139.

35. B. Metz, O. Davidson, P. Bosch, R. Dave, and L. Meyer, eds., *Climate Change 2007: Mitigation of Climate Change* (Cambridge University Press, 2007), p. 244f.

36. J. T. Houghton, *Global Warming* (Cambridge University Press, 2004), p. 250.

37. See *World Energy Outlook 2007* (International Energy Agency, 2007), p. 593.

Index